Regulation of Gene Expression in the Tumor Environment

Menashe Bar-Eli
Editor

Regulation of Gene Expression in the Tumor Environment

Springer

a group of heterogeneous cells that co-exist and interact with an infrastructure of other cells(keratinocytes, fibroblasts, endothelial cells, inflammatory cells) and stromal components, all together known as the tumor microenvironment [1]. The tumor microenvironment is comprised of diverse cell types and elements such as extracellular matrix components (lamin, collagen), growth factors (VEGF, bFGF, thrombin), proteases and interleukins involved in invasion (MMP-2, IL-8, uPA) as well as varying concentrations of oxygen [2]. Furthermore studies have shown that inflammatory cells within the tumor microenvironment contribute to malignancies by releasing growth factors and chemokines [3]. It seems evident that the interaction of tumor cells and the host stroma (microenvironment) is, therefore, essential for tumor progression and, eventually metastasis. Following these same lines, the melanoma tumor microenvironment has emerged within the last decade as a key player in melanocyte transformation and transdifferentiation by providing these necessary elements for growth, invasion and survival [2].

In melanoma, there are several cell types within the tumor microenvironment that influence melanoma progression. For example, keratinocytes, which are found within normal skin, form interactions with melanocytes that are mediated by E-cadherins. Keratinocyte-regulated expression of E-cadherins affects the phenotypic behavior of melanocytes [1]. Disturbances in normal keratinocyte–melanocyte adhesion may contribute to malignant transformation by releasing melanocytes from contact-mediated regulatory controls leading to the advancement of melanoma [4]. Furthermore, keratinocytes induce several pro-angiogenic interleukins (IL-6, IL-8) as well as pro-inflammatory factors (PAF), which may also lead to melanoma progression [3].

Fibroblasts, once thought to play a minimal role in tumorigenesis, have been found to play an important role in potentiating tumor growth. A bi-directional model between melanoma cells and fibroblasts has been proposed in which melanoma cells first produce growth factors such as PDGF, bFGF and TGF-β to activate fibroblasts and endothelial cells and, subsequently, fibroblasts produce a series of growth factors (IGF-1, HGS/SF, bFGF, TGF-β) that further supports the growth and proliferation of melanoma cells [1,4]. These paracrine signaling loops act to create an environmental niche conducive to tumor growth [1].

As can be seen, transformed melanocytic cells will recruit and interact with host cells in the microenvironment. These cells will then become activated and in turn elicit survival, proliferation and invasion signals [4]. The progression of melanoma from radial growth phase to vertical growth phase is accompanied by a myriad of molecular changes that are involved in this transition. Two of the factors involved in this transition that are not only activated by the tumor microenvironment but also in turn affect the microenvironment are the thrombin receptor (PAR-1) and the Platelet Activating Factor Receptor (PAFR).

1.1 PAR-1

Thrombin is a serine protease abundant in the tumor microenvironment milieu, which not only plays a crucial role in blood coagulation but also initiates various cellular responses through the activation of the thrombin receptor, PAR-1 [5]. In fact, activation of coagulation factors have been implicated in tumor growth and are hallmarks of advanced cancers [5,6]. Studies have also demonstrated that tissue factor (TF) is constitutively expressed in melanoma cells and can activate thrombin in a coagulation independent manner, thereby promoting melanoma metastasis [7,8]. In fact, the hypoxic tumor microenvironment also induces TF expression by endothelial cells, tumor associated macrophages and myofibroblasts, thereby also augmenting thrombin in the tumor microenvironment [6].

Furthermore, thrombin-treated tumor cells (including melanoma) enhance their adhesion to platelets and fibronectin *in vitro* [9]. Thrombin also promotes endothelial cell alignment in Matrigel *in vitro* and angiogenesis *in vivo* [10]. It induces the differentiation of endothelial cells into capillary structures in a dose-dependent manner on Matrigel [10]. Furthermore, in the *in vivo* Matrigel system of angiogenesis, there is a 10-fold increase in endothelial cell migration infiltration in response to thrombin. In lung epithelial cells, thrombin was also found to stimulate the expression of PDGF [11]. Blocking of the coagulation pathways at the level of tissue factor, factor Xa, or thrombin, inhibits metastasis of human melanoma cells in SCID mice [8].

Thrombin can also activate several signal transduction pathways through its receptor. The thrombin receptor is a 7-pass transmembrane G-protein coupled receptor. Unlike typical ligand-receptor interactions, thrombin does not activate PAR-1 upon binding. Rather, it cleaves the N-terminus of PAR-1 at serine 42. Upon cleavage, the new amino terminal peptide acts as a tethered ligand that will now bind to the body of the receptor thereby causing cell signaling via G proteins resulting in upregulation of gene products involved in adhesion ($\alpha_{IIb}\beta_3$, $\alpha v\beta_5$, $\alpha_v\beta_3$ integrins) [12–14], invasion (MMP-2) [15], and angiogenesis (IL-8, VEGF, bFGF, PDGF) [11, 16–18]. This suggests that activation of the thrombin receptor may facilitate tumor invasion and metastasis through the induction of cell adhesion molecules, matrix degrading proteases, and stimulating the secretion of angiogenic factors into the melanoma tumor microenvironment, thus contributing to the metastatic phenotype of melanoma.

In human melanoma cells, thrombin acts as a growth factor and is mitogenic, suggesting that signaling by PAR-1 is involved in the biological response of these cells [8]. PAR-1 can also be activated by ligands other than thrombin such as factor Xa, granzyme A, trypsin and plasmin [19–21]. In addition to melanoma, overexpression of PAR-1 has been observed in a variety of human cancers, such as breast, lung, colon, pancreatic and prostate [5, 22–26]. It has also been recently reported that PAR-1 in breast cancer cells can also be proteolytically cleaved and activated by membrane metalloprotease-1 (MMP-1) [27]. Our laboratory has previously demonstrated that PAR-1 is differentially expressed in melanoma cell lines with overexpression being found in highly metastatic cells as compared to nonmetastatic melanoma cell lines [5,28]. Moreover, we found that the overexpression of PAR-1 correlates with the loss of the activator protein-2α (AP-2α), which is a crucial event in the progression of human melanoma [28]. In fact, we observed an inverse correlation between AP-2 and PAR-1 from primary melanoma cell lines

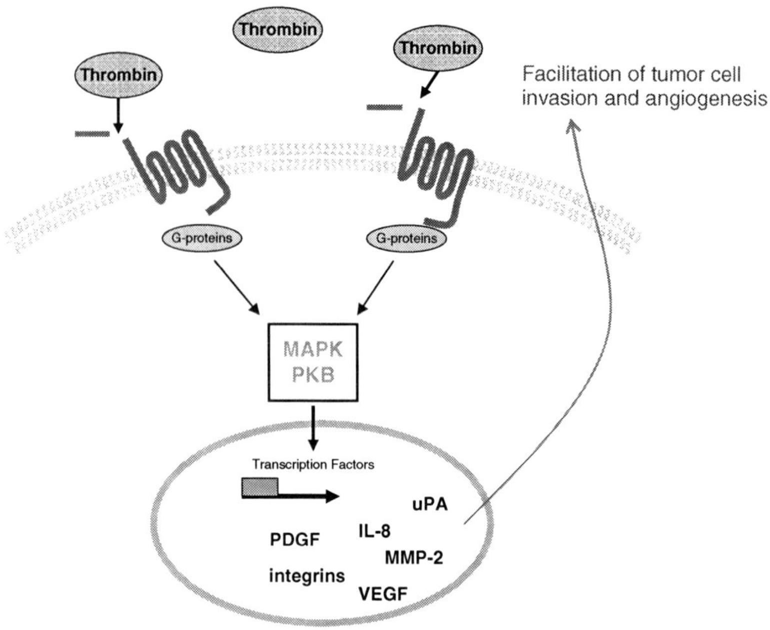

Fig. 1.1 Schematic representation of molecules involved in cell invasion and angiogenesis via activation of PAR-1, which is overexpressed in metastatic melanoma cells. Thrombin from the microenvironment cleaves the N-terminus of PAR-1 to activate the receptor. The tumor-promoting signals transduced by PAR-1 through G-proteins upregulate molecules involved in angiogenesis and invasion

up to highly invasive and aggressive melanomas [28]. Overexpression of PAR-1 is predominantly seen in patients with malignant melanoma tumors and in metastatic lesions as compared to common melanocytic nevi and normal skin [29]. Furthermore, our laboratory has found a significantly higher percentage of PAR-1 positive cells in metastatic melanoma specimens as compared to both dysplastic nevi and primary melanoma specimens [30] attesting to the role of PAR-1 in regulating tumor growth and metastasis of melanoma.

As can be seen, activation of PAR-1 in melanoma cells through different ligands present in the tumor microenvironment will subsequently cause activation of the angiogenic and invasive gene products that are released into the tumor microenvironment (Fig. 1.1). This will also cause activation of fibroblasts and endothelial cells that subsequently forms a more pro-invasive and proliferative environment for melanoma growth and metastasis.

1.2 PAFR

As mentioned previously, it has been shown through genetic and functional experiments that inflammatory cells such as tumor-infiltrating monocytes/macrophages, neutrophils, mast cells, eosinophils, and activated T-lymphocytes contribute to malignancies by the secretion of growth and survival factors, proteases, pro-angiogenic factors and chemokines into the tumor microenvironment [31–34]. In fact, cancer cells promote the recruitment of inflammatory cells, thereby producing inflammatory mediators and angiogenic factors [3].

PAF is secreted into the tumor microenvironment by several cell types, including inflammatory cells, vascular endothelial cells and keratinocytes, which in turn also respond to PAF. Furthermore, platelets in response to thrombin can also secrete PAF. PAF binds and activates the Platelet Activating Factor Receptor (PAFR), a pro-inflammatory mediator, which is also a G-protein coupled receptor. PAFR, in a similar manner to PAR-1, activates signal transduction pathways including MAP kinase, PI3 kinase, PKA and Src pathways [3, 35–39]. Furthermore, our group and others have demonstrated that in human metastatic melanoma cells, PAF can stimulate the activity of p38 MAP kinase [39–41]. PAF activation of these signal transduction pathways results in upregulation of effectors of tumor growth, angiogenesis and malignant progression such as NF-κB, STAT-3 and MMPs [3].

Through the use of PAFR-overexpressing transgenic mice, it was shown that these mice exhibited keratinocyte hyperplasia soon after birth, accompanied by hyperpigmentation, increased melanocytes in ear and tail as well as consequent development of melanoma tumors later in life [42, 43]. These studies also suggested that the recruitment of melanocytes to the dermis was driven by keratinocytes and possibly accumulating fibroblasts and mast cells as the PAFR transgene expression was not seen in melanocytes but was present in keratinocytes. Furthermore the role of PAFR in human melanoma metastasis was further elucidated with *in vivo* experiments using the PAFR antagonist PCA4248. PCA4248 significantly inhibited experimental human melanoma lung metastasis in nude mice [3].

However, it has been shown that PAFR is expressed not only on the surface of keratinocytes but also our lab has shown that all cultured melanoma cell lines regardless of their metastatic potential express constitutively active PAFR [39, 44]. PAFR in melanoma cells is constitutively active in human melanoma cells and mediates gene expression [3].

Our lab also hypothesized that PAFR activation via PAF can phosphorylate and activate the transcription factors cAMP response element-binding (CREB) and activating transcription factor 1 (ATF-1). Expression of these two transcription factors correlate with the transition from radial growth phase to vertical growth phase of human melanoma cells and with their metastatic potential in nude mice [45, 46]. PAF induces CREB and ATF-1 via a PAFR-mediated signal transduction mechanism requiring the Gαq and adenylate cyclase. Furthermore, addition of PAF to the metastatic melanoma A375SM cells stimulated CRE-dependent transcription [39]. Studies have shown that PAF can transactivate membrane type 1-MMP (MT1-MMP) and TIMP-2 genes resulting in proteolytic activation of MMP-2 in human umbilical vein endothelial cells [47]. In human melanoma cells PAF also activated MMP-2 expression and gelatinase activity. Furthermore, MMP-2 activation correlated with an increase in PAF-induced MT1-MMP in human melanoma cells [3, 39].

We propose that all melanoma cells express PAFR regardless of their metastatic potential and secrete basal levels of MMP-2 and MT1-MMP. However, within the melanoma tumor microenvironment where PAF secreting cells such as platelets, endothelial cells and inflammatory cells come into contact with melanoma cells, activation of the PAFR will cause phosphorylation of CREB and ATF-1 through the p38 MAP kinase and PKA signal transduction cascades. Consequently, this results in overexpression and secretion into the microenvironment of MMP-2 and MT1-MMP (Fig. 1.2). However, since only metastatic melanoma cells overexpress CREB

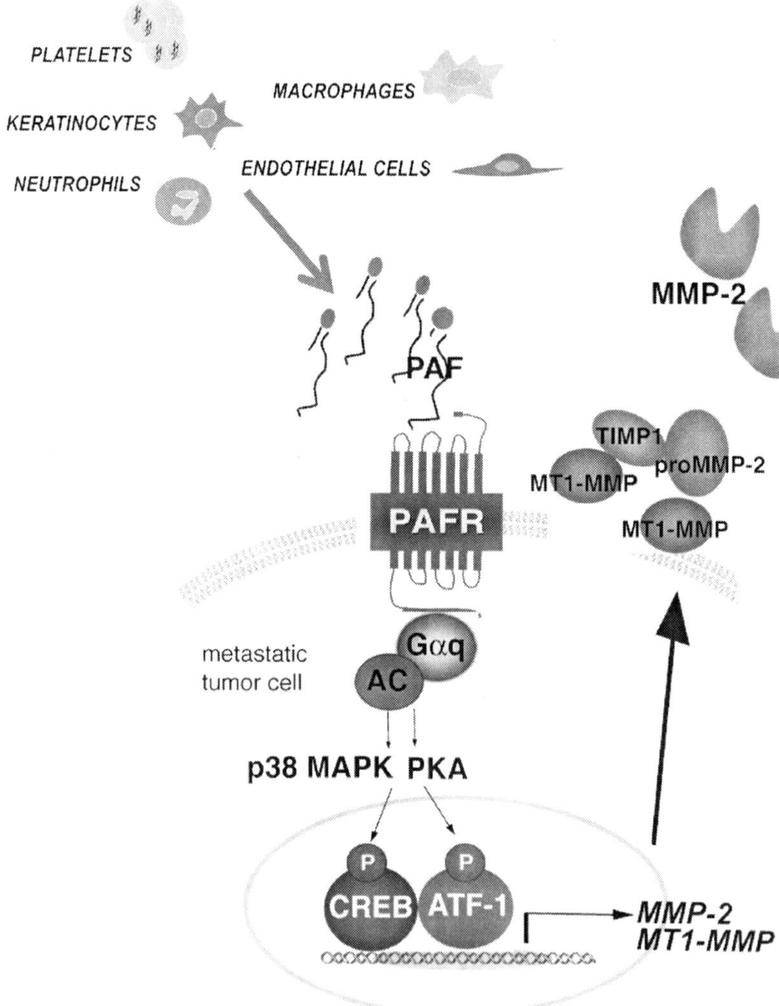

Fig. 1.2 A schematic for the stimulation of MMP-2 and MT1-MMP by PAF via activation of CREB/ATF-1. When melanoma cells come into contact with PAF-producing cells within the tumor microenvironment, PAFR is activated. Through G-proteins and adenylate cyclase, p38 MAPK and PKA phoshporylate CREB and ATF-1. This results in overexpression and secretion of MMP-2 and MT1-MMP

and ATF-1, they are better equipped to respond to the effect of PAF within the tumor microenvironment.

1.3 Conclusion

It is apparent that early inflammatory and angiogenic response and the remodeling of the extracellular proteins are essential factors in creating a microenvironment that sustains tumor growth and metastasis [48]. As we described in this chapter, all these different cell types and factors found within the tumor microenvironment play a significant role in homeostasis and behavior of melanocytes as well as directly affect melanoma growth and malignant invasion [1]. Thrombin, which is abundant in the tumor microenvironment, causes activation of PAR-1, which is found to be upregulated in metastatic melanoma cells. This activation promotes secretion of adhesion, angiogenic and survival factors into the tumor microenvironment allowing for increased metastatic potential of melanoma. Furthermore, PAFR is activated by PAF produced from an array of inflammatory cells, endothelial cells, keratinocytes and platelets found within the tumor microenvironment. Activated PAFR will cause upregulation of the CREB and ATF-1 transcription factors, which in turn increase the secretion of MMP-2 and MT1-MMP. Therefore, melanoma cells will be surrounded by these factors that increase the potential for basement membrane degradation and thereby increase their metastatic potential. Continuing to study the interactions between the tumor microenvironment and melanoma cells will drastically help us understand the mechanisms and key players involved in the transition of human melanoma from radial growth phase to vertical growth phase.

References

1. Lee, J.T., Herlyn, M., *Microenvironmental influences in melanoma progression.* J Cell Biochem 101: 862–872, 2007.
2. Postovit, L.M., Seftor, E.A., Seftor, R.E., Hendrix, M.J., *Influence of the microenvironment on melanoma cell fate determination and phenotype.* Cancer Res 66: 7833–7836, 2006.
3. Melnikova, V., Bar-Eli, M., *Inflammation and melanoma growth and metastasis: the role of platelet-activating factor (PAF) and its receptor.* Cancer Metastasis Rev 26: 359–371, 2007.
4. Hsu, M.Y., Meier, F., Herlyn, M., *Melanoma development and progression: a conspiracy between tumor and host.* Differentiation 70: 522–536, 2002.
5. Tellez, C., Bar-Eli, M., *Role and regulation of the thrombin receptor (PAR-1) in human melanoma.* Oncogene 22: 3130–3137, 2003.
6. Ruf, W., *Tissue factor and PAR signaling in tumor progression.* Thromb Res 120(Suppl 2): S7–S12, 2007.
7. Bromberg, M.E., Konigsberg, W.H., Madison, J.F., Pawashe, A., Garen, A., *Tissue factor promotes melanoma metastasis by a pathway independent of blood coagulation.* Proc Natl Acad Sci USA 92: 8205–8209, 1995.
8. Fischer, E.G., Ruf, W., Mueller, B.M., *Tissue factor-initiated thrombin generation activates the signaling thrombin receptor on malignant melanoma cells.* Cancer Res 55: 1629–1632, 1995.

9. Nierodzik, M.L., Bain, R.M., Liu, L.X., Shivji, M., Takeshita, K., Karpatkin, S., *Presence of the seven transmembrane thrombin receptor on human tumour cells: effect of activation on tumour adhesion to platelets and tumor tyrosine phosphorylation.* Br J Haematol 92: 452–457, 1996.
10. Haralabopoulos, G.C., Grant, D.S., Kleinman, H.K., Maragoudakis, M.E., *Thrombin promotes endothelial cell alignment in Matrigel in vitro and angiogenesis in vivo.* Am J Physiol 273: C239–C245, 1997.
11. Shimizu, S., Gabazza, E.C., Hayashi, T., Ido, M., Adachi, Y., Suzuki, K., *Thrombin stimulates the expression of PDGF in lung epithelial cells.* Am J Physiol Lung Cell Mol Physiol 279: L503–L510, 2000.
12. Even-Ram, S.C., Maoz, M., Pokroy, E., Reich, R., Katz, B.Z., Gutwein, P., Altevogt, P., Bar-Shavit, R., *Tumor cell invasion is promoted by activation of protease activated receptor-1 in cooperation with the alpha vbeta 5 integrin.* J Biol Chem 276: 10952–10962, 2001.
13. Wojtukiewicz, M.Z., Tang, D.G., Nelson, K.K., Walz, D.A., Diglio, C.A., Honn, K.V., *Thrombin enhances tumor cell adhesive and metastatic properties via increased alpha IIb beta 3 expression on the cell surface.* Thromb Res 68: 233–245, 1992.
14. Senger, D.R., Ledbetter, S.R., Claffey, K.P., Papadopoulos-Sergiou, A., Peruzzi, C.A., Detmar, M., *Stimulation of endothelial cell migration by vascular permeability factor/vascular endothelial growth factor through cooperative mechanisms involving the alphavbeta3 integrin, osteopontin, and thrombin.* Am J Pathol 149: 293–305, 1996.
15. Zucker, S., Conner, C., DiMassmo, B.I., Ende, H., Drews, M., Seiki, M., Bahou, W.F., *Thrombin induces the activation of progelatinase A in vascular endothelial cells: physiologic regulation of angiogenesis.* J Biol Chem 270: 23730–23738, 1995.
16. Ueno, A., Murakami, K., Yamanouchi, K., Watanabe, M., Kondo, T., *Thrombin stimulates production of interleukin-8 in human umbilical vein endothelial cells.* Immunology 88: 76–81, 1996.
17. Huang, Y.Q., Li, J.J., Hu, L., Lee, M., Karpatkin, S., *Thrombin induces increased expression and secretion of VEGF from human FS4 fibroblasts, DU145 prostate cells and CHRF megakaryocytes.* Thromb Haemost 86: 1094–1098, 2001.
18. Cucina, A., Borrelli, V., Di Carlo, A., Pagliei, S., Corvino, V., Santoro-D'Angelo, L., Cavallaro, A., Sterpetti, A.V., *Thrombin induces production of growth factors from aortic smooth muscle cells.* J Surg Res 82: 61–66, 1999.
19. O'Brien, P.J., Molino, M., Kahn, M., Brass, L.F., *Protease activated receptors: theme and variations.* Oncogene 20: 1570–1581, 2001.
20. Ruf, W., Mueller, B.M., *Thrombin generation and the pathogenesis of cancer.* Semin Thromb Hemost 32(Suppl 1): 61–68, 2006.
21. Hansen, K.K., Saifeddine, M., Hollenberg, M.D., *Tethered ligand-derived peptides of proteinase-activated receptor 3 (PAR3) activate PAR1 and PAR2 in Jurkat T cells.* Immunology 112: 183–190, 2004.
22. Even-Ram, S., Uziely, B., Cohen, P., Grisaru-Granovsky, S., Maoz, M., Ginzburg, Y., Reich, R., Vlodavsky, I., Bar-Shavit, R., *Thrombin receptor overexpression in malignant and physiological invasion processes.* Nat Med 4: 909–914, 1998.
23. Henrikson, K.P., Salazar, S.L., Fenton, J.W., II, Pentecost, B.T., *Role of thrombin receptor in breast cancer invasiveness.* Br J Cancer 79: 401–406, 1999.
24. Rudroff, C., Schafberg, H., Nowak, G., Weinel, R., Scheele, J., Kaufmann, R., *Characterization of functional thrombin receptors in human pancreatic tumor cells (MIA PACA-2).* Pancreas 16: 189–194, 1998.
25. Wojtukiewicz, M.Z., Tang, D.G., Ben-Josef, E., Renaud, C., Walz, D.A., Honn, K.V., *Solid tumor cells express functional "tethered ligand" thrombin receptor.* Cancer Res 55: 698–704, 1995.
26. Kaushal, V., Kohli, M., Dennis, R.A., Siegel, E.R., Chiles, W.W., Mukunyadzi, P., *Thrombin receptor expression is upregulated in prostate cancer.* Prostate 66: 273–282, 2006.

27. Boire, A., Covic, L., Agarwal, A., Jacques, S., Sherifi, S., Kuliopulos, A., *PAR1 is a matrix metalloprotease-1 receptor that promotes invasion and tumorigenesis of breast cancer cells.* Cell 120: 303–313, 2005.
28. Tellez, C., McCarty, M., Ruiz, M., Bar-Eli, M., *Loss of activator protein-2alpha results in overexpression of protease-activated receptor-1 and correlates with the malignant phenotype of human melanoma.* J Biol Chem 278: 46632–46642, 2003.
29. Massi, D., Naldini, A., Ardinghi, C., Carraro, F., Franchi, A., Paglierani, M., Tarantini, F., Ketabchi, S., Cirino, G., Hollenberg, M.D., Geppetti, P., Santucci, M., *Expression of protease-activated receptors 1 and 2 in melanocytic nevi and malignant melanoma.* Hum Pathol 36: 676–685, 2005.
30. Tellez, C.S., Davis, D.W., Prieto, V.G., Gershenwald, J.E., Johnson, M.M., McCarty, M.F., Bar-Eli, M., *Quantitative analysis of melanocytic tissue array reveals inverse correlation between activator protein-2alpha and protease-activated receptor-1 expression during melanoma progression.* J Invest Dermatol 127: 387–393, 2007.
31. Balkwill, F., Mantovani, A., *Inflammation and cancer: back to Virchow?* Lancet 357: 539–545, 2001.
32. Coussens, L.M., Werb, Z., *Inflammation and cancer.* Nature 420: 860–867, 2002.
33. Mantovani, A., *Cancer: inflammation by remote control.* Nature 435: 752–753, 2005.
34. Pikarsky, E., Porat, R.M., Stein, I., Abramovitch, R., Amit, S., Kasem, S., Gutkovich-Pyest, E., Urieli-Shoval, S., Galun, E., Ben-Neriah, Y., *NF-kappaB functions as a tumour promoter in inflammation-associated cancer.* Nature 431: 461–466, 2004.
35. Coffer, P.J., Schweizer, R.C., Dubois, G.R., Maikoe, T., Lammers, J.W., Koenderman, L., *Analysis of signal transduction pathways in human eosinophils activated by chemoattractants and the T-helper 2-derived cytokines interleukin-4 and interleukin-5.* Blood 91: 2547–2557, 1998.
36. Franklin, R.A., Mazer, B., Sawami, H., Mills, G.B., Terada, N., Lucas, J.J., Gelfand, E.W., *Platelet-activating factor triggers the phosphorylation and activation of MAP-2 kinase and S6 peptide kinase activity in human B cell lines.* J Immunol 151: 1802–1810, 1993.
37. Honda, Z., Takano, T., Gotoh, Y., Nishida, E., Ito, K., Shimizu, T., *Transfected platelet-activating factor receptor activates mitogen-activated protein (MAP) kinase and MAP kinase kinase in Chinese hamster ovary cells.* J Biol Chem 269: 2307–2315, 1994.
38. Ishii, S., Nagase, T., Shimizu, T., *Platelet-activating factor receptor.* Prostaglandins Other Lipid Mediat 68–69: 599–609, 2002.
39. Melnikova, V.O., Mourad-Zeidan, A.A., Lev, D.C., Bar-Eli, M., *Platelet-activating factor mediates MMP-2 expression and activation via phosphorylation of cAMP-response element-binding protein and contributes to melanoma metastasis.* J Biol Chem 281: 2911–2922, 2006.
40. Landis, M., Yi, Q., Hyatt, A.M., Travers, A.R., Lewis, D.A., Travers, J.B., *Involvement of P38 MAP kinase in the augmentation of UVB-mediated apoptosis via the epidermal platelet-activating factor receptor.* Arch Dermatol Res 299: 263–266, 2007.
41. Nick, J.A., Avdi, N.J., Young, S.K., Knall, C., Gerwins, P., Johnson, G.L., Worthen, G.S., *Common and distinct intracellular signaling pathways in human neutrophils utilized by platelet activating factor and FMLP.* J Clin Invest 99: 975–986, 1997.
42. Ishii, S., Nagase, T., Tashiro, F., Ikuta, K., Sato, S., Waga, I., Kume, K., Miyazaki, J., Shimizu, T., *Bronchial hyperreactivity, increased endotoxin lethality and melanocytic tumorigenesis in transgenic mice overexpressing platelet-activating factor receptor.* Embo J 16: 133–142, 1997.
43. Sato, S., Kume, K., Ito, C., Ishii, S., Shimizu, T., *Accelerated proliferation of epidermal keratinocytes by the transgenic expression of the platelet-activating factor receptor.* Arch Dermatol Res 291: 614–621, 1999.
44. Travers, J.B., Huff, J.C., Rola-Pleszczynski, M., Gelfand, E.W., Morelli, J.G., Murphy, R.C., *Identification of functional platelet-activating factor receptors on human keratinocytes.* J Invest Dermatol 105: 816–823, 1995.

45. Darst, M., Al-Hassani, M., Li, T., Yi, Q., Travers, J.M., Lewis, D.A., Travers, J.B., *Augmentation of chemotherapy-induced cytokine production by expression of the platelet-activating factor receptor in a human epithelial carcinoma cell line.* J Immunol 172: 6330–6335, 2004.
46. Rutberg, S.E., Goldstein, I.M., Yang, Y.M., Stackpole, C.W., Ronai, Z., *Expression and transcriptional activity of AP-1, CRE, and URE binding proteins in B16 mouse melanoma subclones.* Mol Carcinog 10: 82–87, 1994.
47. Axelrad, T.W., Deo, D.D., Ottino, P., Van Kirk, J., Bazan, N.G., Bazan, H.E., Hunt, J.D., *Platelet-activating factor (PAF) induces activation of matrix metalloproteinase 2 activity and vascular endothelial cell invasion and migration.* Faseb J 18: 568–570, 2004.
48. van Kempen, L.C., van Muijen, G.N., Ruiter, D.J., *Stromal responses in human primary melanoma of the skin.* Front Biosci 10: 2922–2931, 2005.

Chapter 2
Functions of Autocrine Motility Factor at the Tumor Microenvironment

Tatsuyoshi Funasaka and Avraham Raz

Abstract Autocrine motility factor (AMF) is a tumor-secreted cytokine and is abundant at tumor sites, where it may affect the process of tumor growth and metastasis. AMF is a multifunctional protein capable of affecting cell migration, invasion, proliferation, and survival, and possesses phosphoglucose isomerase activity and can catalyze the step in glycolysis and gluconeogenesis. Here, we review the role of AMF and tumor environment on malignant processes. The outcome of metastasis depends on multiple interactions between tumor cells and homeostatic mechanisms; therefore elucidation of the tumor/host interactions in the tumor microenvironment is essential in the development of new prevention and treatment strategies. Such knowledge might provide clues to develop new future therapeutic approaches for human cancers.

Keywords Motility factor · Autocrine effect · Survival · Apoptosis · Tumor-host interaction · Extracellular matrix · Collagen · Cytokines

2.1 Introduction

For patients diagnosed with cancer, the prognosis of the disease hinges on whether metastasis develops or not. Most deaths from cancer are due to metastases that are resistant to conventional therapies despite significant improvements in diagnosis, surgical techniques, patient care, and adjuvant therapies. It occurs when tumor cells break away from the primary lesion and lodge in tissue far removed from the organ where the disease first developed. There is an urgent need to develop effective strategies for the treatment of metastatic tumors.

To develop the way for cancer therapy from a biological point of view, it is necessary to analyze the sequence of events known as the "metastatic cascade". Before metastasis occurs, malign neoplastic are made up of cell populations exhibiting a

A. Raz
Tumor Progression and Metastasis Program, Barbara Ann Karmanos Cancer Institute, Wayne State University, School of Medicine, Detroit, MI 48201, USA
e-mail: raza@karmanos.org

M. Bar-Eli (ed.), *Regulation of Gene Expression in the Tumor Environment*,

wide range of biological heterogeneity with regard to properties such as cell surface, proliferative index, antigenicity, immunogeicity, and expression of phenotypic qualities enabling them to invade other tissues [1, 2]. Invasion is the initiating event in the metastatic cascade. Proteolytic enzymes (lysosomal hydrolases, collagenases) secreted by tumor cells degrade basement membrane constituents such as type IV collagen, laminin, and fibronectin, allowing invading cell access to the underlying connective tissue matrix [3–5]. Degradation and movement through this matrix is the next step in the metastatic cascade, before the cell ultimately invades a vascular endothelial basement membrane to enter an adjacent blood vessel, lymphatic channel, or nerve [6, 7]. Metastasis occurs in most cases via the blood and/or lymph system. Early clinical observations suggested that solid tumors (carcinomas) spread primarily via the lymphatic vessels and that mesenchymal (connective tissue) tumors spread mainly through the bloodstream. In truth, the lymphatic and vascular systems have numerous connections that allow disseminating tumor cells to pass rapidly from one system to the other [8]. Entering into these system by tumor cells is facilitated by the structure of the microcirculation, especially in new capillary blood vessels with fenestrated or discontinuous endothelium, a lack of stable intercellular junctions between the endothelial cells and discontinuous or absent basement membrane [8, 9]. Once the tumor cells have made their way into microcirculation, they are carried by the flow to distant organs. The invading tumor cell must survive natural host immunity (macrophages, NK cells and cytotoxic T lymphocytes) [10] and passive mechanical forces [11] in these vascular compartments in order to enter a distant organ system (secondary invasion). Once the tumor cells have adhered to the microvascular endothelium, the extravasation of the tumor cells into the organ begins (migration out of blood vessels) [12]. After passing the endothelial barrier, the tumor cells produce enzymes that break down the components of the basement membrane and underlying connective tissue as mentioned above, thus facilitating their passage into the parenchyma of organ. Finally, tumor cells continue to proliferate in the target organ, which depends on establishing an adequate blood supply (angiogenesis), and form metastatic foci [2, 13, 14]. Not every tumor cell that is endowed with metastatic potential survives to proliferate at a distant site. It has been estimated that fewer than 2% of cells in a tumor mass have undergone the phenotypic changes necessary for metastasis [15], and indeed, less than 0.1% of cells in a tumor mass survive the entire cascade of events to proliferate into a metastatic focus [16].

The major obstacle to the treatment of metastases is the biological heterogeneity of tumor cells in primary and secondary tumors. This heterogeneity is exhibited in a wide range of genetic, biochemical, immunological, and biological characteristics including cell morphologies, growth properties, and ability to invade [1, 2]. Drug susceptibility differences between metastatic lesions and their primary tumors are well documented [17]. This infers adaptation and differential gene expression between the primary and metastasis tumor. These differences are believed to result from selective genetic changes [18]. Aspects of the host microenvironment have been shown to affect the genes that regulate metastasis [19–21].

2.2 The Tumor Microenvironment

The influence of the organ microenvironment on tumor biology has been recognized for many years, since Paget's "seed and soil" hypothesis [22]. According to the "seed and soil" theory, a tumor cell capacity for metastatic colonization is determined by its special characteristics ("seed") and by the host organ that tumor cells encounter as they travel ("soil"), suggesting that progression of metastases is not random. Metastasis resulted only when the "seed and soil" were compatible. Isaiah J. Fidler defines the current version of the "seed and soil" hypothesis as having two principles [23]: first, neoplasms are heterogeneous and consist of cells with different biologic properties; second, the outcome of cancer growth and spread depends on multiple interactions of tumor cells with host homeostatic factors. Both the "soil" and the "soil" profoundly influence the outcome of systemic therapy for cancer.

There have been many reports that indicate the organ microenvironment can influence the biology of cancer growth and metastasis in several ways. To gain entrance into the microcirculation, tumor cells must degrade connective tissue, extracellular matrix, and basement membrane components that constitute barriers against invading tumor cells [24]. Metastatic tumor cells produce various proteases and glycosidases that degrade extracellular matrix, and the production of such enzymes directly correlates with invasion and formation of metastasis [24, 25]. The expression level and activity of collagenase type IV in human carcinoma cells are influenced by cytokines produced by specific tissues and organs [25]. Once tumor cells reach the parenchyma of distant organs, they must proliferate to establish a metastasis. To do so, metastatic cells can seize physiologic growth factors that are produced by the microenvironment [20, 21]. Furthermore, molecular modifications in adhesion molecule, cytokine, or growth factor expression of metastatic tumor cells have been demonstrated to change depending on the tumor location *in vivo* [19, 26]. These observations suggest that host tissue environment can regulate the differential expression of tumor cell proteins. Additionally, tumor-secreted cytokines/growth factors modify the local environment surrounding the tumor to modulate the immune response, inhibit vascular cell adhesion protein expression, and induce angiogenesis [18,27,28].

The tumor microenvironment regulates the expression of tumor-specific and organ-specific factors. Tumor cells are capable of secreting various cytokines that affect both the different cells of the host (endothelial cells, fibroblasts, mast cells, macrophages and the cells of the parenchyma itself) and extracellular matrix, which in turn secrete different factors that reciprocally affect the metastatic cells [29–32]. One of the most important cytokine secreted by tumor cells is autocrine motility factor (AMF), originally purified from the conditioned medium of human A2058 melanoma cells as a tumor producing cytokine that stimulates both direct and random migration [33].

Cell migration is essential for development, inflammation and tissue repair, but it also allows malignant cells to exert their lethal ability to invade tissues and metastasize [6, 7]. Malignant tumors are characterized by their unrestrained growth and invasion into surrounding host tissue. Pathologic observations have clearly demonstrated that invasion is not only a consequence of tumor growth compression but also involves the dynamic locomotion of tumor cells through host tissue barriers [34]. Individual tumor cells or small groups of tumor cells migrating away from the main mass often are found at the invasion front in the observation of histological sections of malignant tumors. The metastatic dissemination of tumor cells to distant organs requires highly motile behavior during both entrance and exit from blood vessels. Tumor cell locomotion is regulated by several secreted cytokines having cellular kinesis induction properties in an autocrine and/or paracrine manner [15, 16]. Host serum proteins and breakdown products of extracellular matrix can influence cell motility as well [35, 36]. AMF is one of the major cell motility-stimulating factors associated with the development of tumors.

2.3 Autocrine Motility Factor

AMF was originally purified from the conditioned medium of human A2058 melanoma cells as a tumor producing cytokine with an estimated size of 55 kDa that stimulates both direct and random migration [33]. In clinical cancer pathology, the presence of AMF in the serum and urine is of prognostic value indicating cancer progression [37–39]. Moreover, over-expression of AMF gene in non-tumorigenic cells induces AMF secretion and leads to a gain of tumorigenic properties *in vivo* [40, 41].

The remarkable property of AMF is identified with glycolytic enzyme, phosphoglucose isomerase (PGI) [42]. PGI is the second enzyme in the glycolytic pathway and catalyzes the interconversion of glucose 6-phosphate and fructose 6-phosphate during glycolysis and gluconeogenesis. Thus, it is highly conserved in bacteria and eukaryotes [43]. Mutations in PGI are a cause of hereditary nonspherocytic hemolytic anemia, and PGI deficiency can be associated with hydrops fetalis and immediate neonatal death [44–46], and PGI is an antigen of arthritis disease [47]. Accordingly, PGI plays an important role in the cellular metabolism. Molecular cloning and sequencing have identified PGI as an AMF, and it is recognized that AMF has PGI enzymatic activity, AMF activities are inhibited by specific PGI inhibitors including erythrose 4-phosphate (E4P) and mannose 6-phosphate (carbohydrate phosphates) [42], and commercial PGI has AMF activities. Therefore, inside the cell PGI functions in glucose metabolism, while outside the cell it acts as a cytokine. As mentioned above, there have been many reports that PGI/AMF is involved in tumor metastasis and invasion, and PGI/AMF is associated with several malignant conditions, such as induction of angiogenesis [48]. It seems that tumor cells need glycolysis more than normal cells; up regulation of glycolytic metabolism occurs downstream of multiple oncogenic pathways and has been shown to correlate with increased tumor aggressiveness and poor patient prognosis in several tumor types [43, 49]. These observations suggest that the glycolytic phenotype plays a role in tumor progression by contributing to tumor growth or survival. Furthermore, AMF exhibits the ability to function as a neuroleukin promoting growth of embryonic spinal and sensory neurons [50], as a maturation factor mediating differentiation of human myeloid leukemia cells [51], as a sperm antigen-36 [52], or as a myofibril-bound serine proteinase inhibitor [53]. Although the peptide sequence of AMF suggested homology [54], many reports indicate that AMF, PGI, neuroleukin, maturation factor, sperm antigen-36, and myofibril-bound serine proteinase inhibitor have different functions: AMF/PGI/neuroleukin/maturation factor/sperm antigen-36/myofibril-bound serine proteinase inhibitor is a multifunctional protein.

The motility stimulation with AMF is mediated by its interaction with receptors on the surface of target cells [55]. AMF receptor (AMFR/gp78) is a 78 kDa seven-transmembrane glycoprotein which belongs to the G protein coupled receptor family [55, 56]. There have been many reports that over expression of AMF and AMFR are correlated with progression of malignant tumors. The levels of PGI/AMF and its cell surface receptor AMFR expressions are associated with the pathologic stage, grade, and degree of tumor penetration to surrounding tissues marking a poor prognosis [57–61].

2.4 The Role of AMF in Tumor Angiogenesis

Metastasis is an important clinical parameter in the prognosis of patients who develop malignant tumors. Abundant studies have suggested that metastasis is one of the most complicated biological phenomenon and enormous factors are interacting between tumor and host during tumor invasiveness and metastasis. It generally occurs via the vascular or lymphangial system on distant organs such as the liver or lung metastasis of colon cancer; therefore, metastasis is closely related to the vascular system. Another important relation between metastasis and blood vessels is angiogenesis. Angiogenesis, formation of capillary blood vessels leading to neovascularization, is an organic reaction caused by endothelial cell growth and migration from pre-existing blood vessels [62], and is essential to form a circulatory system or fabric at the embryonic phase in vertebrate [63]. It is also associated with an array of pathologic processes including inflammatory disease, diabetic retinopathy and wound-healing [64]. Capillary blood vessels are arranged in a series of cellular processes, that is, endothelial cells can migrate, propagate, organize to lumen and form new capillaries in response to appropriate angiogenic signals [64]. A number of angiogenic factors have been identified, such as vascular endothelial growth factor (VEGF), basic-, acidic-fibroblast growth factor (bFGF, aFGF), platelet-derived endothelial cell growth factor (PD-ECGF), transforming growth factor ß (TGF-ß), tumor necrosis factor (TNF), angiogenin and interleukin 8 (IL-8) [62, 64, 65]. For solid tumors of more than several millimeters in diameter, nutrition and oxygen supplies are essential from tumor-generated new blood vessels. Therefore, solid tumors cannot grow without the induction of angiogenesis [66]. Tumor growth is accelerated with the induction of angiogenesis, invading into the surrounding host tissue, and disseminating to distant organs. There are reported to be a correlation between tumor vascular density and clinical malignancy in numerous malignant tumors, and an unfavorable prognosis for tumors with a high density of vessels [66–68].

It is well known that malignant tumor cells secrete various angiogenic factors. Extensive research has shown that the VEGF-system plays an important part in most cases as a tumor angiogenic factor, and other investigations suggested that some cytokines like PD-ECGF or IL-8 are closely related to cancerous angiogenesis [69, 70]. It is thought that the regulation mechanisms of these various angiogenic cytokines are also modulated or controlled by the surrounding extracellular environment, and the complexities of the angiogenic phenomena and physiologic angiogenesis are due to expression and interaction of coordinated multiple cytokines.

Molecular mechanisms of tumor angiogenesis have been investigated since angiogenesis was recognized to play an important role in solid tumor progression. It is well known that the processes of neovascularization are postulated to synchronize with the up-regulation or down-regulation of several angiogenic factors [65]. Tumor angiogenesis is promoted by angiogenic-stimulating factors alone or in combination, such as VEGF, PD-ECGF, and IL-8 which are over expressed in solid tumors [65, 66, 71]. The angiogenic factors modulate the single or multiple phases among the three processes of angiogenesis: (i) enzymatic degradation of the basement membrane; (ii) endothelial cell migration; and (iii) endothelial cell multiplication. The mechanisms of angiogenesis have been elucidated with the demonstration of the roles of factors in each process. Thus, we examined whether AMF may exhibit angiogenic activity by focusing on endothelial cell motility [48].

The effect of AMF on endothelial cell motility was evaluated by phagokinetic analysis of HUVECs, and the motility response was stimulated approximately 2.0-fold at an AMF concentration of 50 pg/ml. AMFR was barely detected on the surface of untreated cells, while AMF-exposed cells exhibited intensive AMFR expression, which was localized predominantly in a single perinuclear pattern. This expression of AMFR was closely correlated with their motile response. To accomplish the process of angiogenesis, migrating endothelial cells must undergo morphogenesis such as formation of capillary-like tubes. AMF induced a network of branched and associated elongated cells which often anastomosed with one another when HUVECs were cultured on collagen gels. Moreover, the angiogenic activity of AMF was evaluated by *in vivo* assays. AMF could induce angiogenesis in Matrigel plugs *in vivo* [72], AMF-over expressed tumor cell stable transfectants induced the development of many new capillary blood vessels compared to mock cells, and the AMF-induced angiogenesis was inhibited by specific AMF inhibitor E4P. Therefore, we have come to the conclusion that AMF is a tumor-derived angiogenic factor for the surrounding tissues. AMF had been recognized as a type of autocrine-type cytokine that acts only against tumor cells as the name implies, however, it is suggested that AMF can affect normal surrounding tissues in a paracrine manner. In short, AMF plays an essential role and contributes substantially to tumor angiogenesis by its motile stimulation activity.

2.5 AMF/VEGF Crosstalk During Tumor Progression

The growth of solid tumors and the formation of metastasis are dependent on angiogenesis [66]. Angiogenesis is a complex, multi-step process that results in the formation of new blood vessels from preexisting vasculature [62]. During tumor angiogenesis, endothelial cells degrade the basement membrane by releasing enzymes, migrate through the membrane, and finally proliferate to extend the blood vessel into the tumor. Many autocrine or paracrine factors such as bFGF, aFGF, TGF-ß, TNF, VEGF, and their receptors are needed as either stimulators or inhibitors for the various steps involved in this complex process [62, 64–66]. Among the various factors mentioned above, VEGF is considered to be the prime regulator of angiogenesis, vasculogenesis and vascular permeability [73–75]. VEGF is expressed by almost all solid tumors.

VEGF acts through two high-affinity, tyrosine-phosphorylating, transmembrane receptors (VEGFR) named Flt-1 and KDR, which have been identified almost specifically on human endothelial cells [76, 77]. The two VEGF receptors have been shown to be expressed preferentially in the proliferating endothelium of vessels lining and penetrating solid tumors, whereas they are almost undetectable in the vessels of healthy tissue [78, 79]. Thus, expression of both VEGF and its receptors seems to be a prerequisite for tumor angiogenesis. Interference with the VEGF-VEGFR signaling system has been shown to significantly inhibit tumor growth and metastasis [69, 70, 80].

Tumor angiogenesis and ascites accumulation are complicated phenomena caused by many factors produced by interacting between neoplasms and hosts. It is also considered that there are varied interactions among those factors. So we hypothesized the signal crosstalk between VEGF-VEGFR and AMF-AMFR system as a clue of this complicated phenomenon [81].

Flt-1 or KDR regulation on the host endothelial cells is significant in the case of tumor angiogenesis and metastasis. AMF stimulates Flt-1 expression on HUVECs in a dose- and time-dependent manner; however, AMF does not affect the expression of KDR. The biological responses to VEGF in AMF-pretreated HUVECs, including cell motility, were higher than those of untreated cells. It has been reported that proliferative signals of VEGF in endothelial cells are mainly dependent on KDR; on the other hand, migrational activities are dependent on Flt-1 [82, 83]. AMF activates PKC and phosphotidylinositol 3′-kinase (PI3K) in endothelial cells leading to increased Flt-1 expression in a paracrine manner. Furthermore, increased Flt-1 expressions were found on the newly developed blood vessels of the mouse tissue exposed to AMF. Flt-1-positive cells were arranged on the blood vessel wall. Therefore, AMF stimulates Flt-1 expression in a physiological condition and contributes secondarily to VEGF activity such as potent mitogenic effect for endothelial cells and induction of angiogenesis.

AMF induces tumor angiogenesis *in vitro* as well as *in vivo* by motile stimulating effect to endothelial cells. Furthermore, AMF induces the activation of PKC and PI3K leading to increasing Flt-1 expression in a paracrine manner, which brings about an increased biological responsiveness to VEGF in endothelial cells.

In addition, VEGF can increase Flt-1 expression by itself [84]. This malignant cycle comprising of AMF and VEGF will result in marked locomotive, angiogenic and further metastatic synergy [80, 82].

2.6 The AMF Expression in Hypoxia

A reduction in tissue oxygen levels is called "hypoxia". The functional definition of hypoxia is insufficient delivery of oxygen for the demand of the tissue [85]. This supply/demand mismatch can occur if there is an increase in demand for oxygen that blood flow cannot meet, as in the exercising muscle, or when there is a reduction in the delivery of oxygen, as in the vascular insufficiency that can occur in coronary artery disease. Therefore, it is apparent that cellular hypoxia can arise from physiologic circumstances such as exercise or travel to high elevations, as well as from pathophysiologic conditions such as poorly formed tumor vasculature. Hypoxia is seen in several pathophysiologic processes including ischemia, pulmonary diseases and cancer [85]. Tumor hypoxia is an important indicator of cancer prognosis; it contributes to tumor progression and poor response to radiotherapy and chemotherapy [86]. Aggressive tumors often have insufficient blood supply because of their rapid expansion. Hypoxic stress in solid tumors is known to lead new blood vessel formation, known as angiogenesis or neovascularization, to supply oxygen to starved tissues [13]. Cellular adaptation to hypoxia and tumor neovascularization is associated with the ability of invasion and metastasis during further growth of the primary tumor [49]. Other several cellular responses including resistance to apoptosis, erythropoiesis and glycolysis are activated in hypoxia [86, 87]. A key regulator of these cellular responses to oxygen deficiency is the transcriptional factor, hypoxia inducible factor 1 (HIF-1).

HIF-1 is a transcriptional factor that regulates changes in gene expression in response to changes in cellular oxygen concentrations. HIF-1 is a heterodimer composed of two subunits, an oxygen-regulated HIF-1α that determines HIF activity and a constitutively expressed HIF-1ß (also known as aryl hydrocarbon receptor nuclear translocator or ARNT) [88]. Under normoxic conditions (21% O_2, or 158 mm Hg partial pressure), HIF-1α protein is extremely unstable and quickly degraded through the ubiquitin-proteasome pathway although HIF-1α gene is continuously expressed [89, 90]. Recently, it has been shown that the tumor suppressor von Hippel-Lindau protein (pVHL) binds to specific hydroxylated HIF-1α on proline residues upon normoxia and leads HIF-1α to be targeted to the proteasome [91–93]. Under hypoxic conditions, the hydroxylases stop functioning and HIF-1α escapes the degradation. Consequently, stabilized HIF-α translocates to the nucleus, dimerizes with a HIF-1ß subunit and the heterodimer then binds to hypoxia response elements (HREs) in the promoters and enhancers of target genes. HIF-1 activates the transcription of a large number of genes whose protein products are critical for tumor progression including angiogenesis, which regulates by angiogenic factor such as VEGF, metabolic adaptation (glucose transporters and glycolytic pathway enzymes) and apoptosis resistance (endothelin-1, insulin-like growth factor 2, and TGF-α) [87]. Several HIF-responsive genes are known to be over expressed in human malignancies [49].

HIF-1 plays a central role in oxygen homeostasis by inducing the expression of a broad range of genes in a hypoxic-dependent manner, including VEGF, Flt-1, inducible NO synthase, transferrin, and many genes in the glycolytic pathway such as glucose transporter 1, hexokinase, phosphofructokinase, glyceraldehydes-3-phosphate dehydrogenase, and phosphoglycerate kinase [49, 87, 94–96]. There has been reported that hypoxia is an inducer of AMF expression, and AMF is one of those HIF-1-inducible genes [97–99]. The expression of AMF mRNA/protein is up-regulated in some tumor cells by hypoxia and hypoxia-induced AMF expression is mediated by HIF-1 [100]. In addition, the hypoxic induction of AMF expression is suppressed by inhibitors of VEGF or VEGF receptors, suggesting that hypoxia-inducible VEGF regulates the AMF expression. Hypoxia also enhanced tumor cell motility, and these effects were strongly inhibited by the AMF, VEGF, or VEGF receptor inhibitors [100]. A principal mediator of tumor angiogenesis is VEGF; a major transcriptional activator of the VEGF gene is HIF-1 [87]. Thus, it was reasonable to assume that the expressions of VEGF and AMF under hypoxia might be related. Hypoxic induction of AMF could be regulated at least in part by VEGF.

2.7 Autocrine/Paracrine Function of AMF in Cancer

One of the most important characteristics of tumor cells is their ability to grow in unusual locations, especially at metastatic sites. The successful proliferation of tumor cells is due to their responses to local (paracrine; be made and secreted by one cell and act on adjacent cells in a tissue or organ) growth factors and inhibitors and their production and responses to their own (autocrine; be made by and act on the same cell) growth factors [30]. As tumors grow and develop, they undergo changes in their growth and other properties. For example, when tumor cells invade and spread to other sites at the early stages of malignant tumor progression, there is a tendency for many common cancers to metastasize and grow preferentially at particular sites, suggesting that unique tissue paracrine growth mechanisms may dominate the growth signals processed by metastatic cells. At somewhat later stages of tumor progression, where widespread dissemination to various tissues and organs occurs, autocrine growth mechanisms may dominate. The progression of malignant cells to completely autonomous growth states can occur, and at this stage of tumor progression cell proliferation may be independent of growth factors or inhibitors [30–32].

Autocrine regulation of tumor cells involves the endogenous production of growth factors which act on the producer cells on specific receptors and stimulate cell proliferation. The essential requirements for a growth factor to function in an autocrine manner are as follows: (i) the growth factor protein is produced; (ii) the growth factor receptor is expressed [29–32]. Some growth factors produced by tumor cells can have paracrine effects on other cells in the surrounding tissues. These exogenous paracrine factors either enhance tumor growth, angiogenesis, adhesion or motility, promote dissemination or cause differentiation [31,32].

There is now increasing evidence supporting a role for the AMF/AMFR axis in cancer, possibly via autocrine/paracrine mechanisms. The co-expression of AMF and the AMFR in human cancers contribute to their development [57, 58]. Autocrine effects of AMF involve the stimulation of tumor cell motility and invasion, the influence of tumor proliferation, and inducing resistant to apoptosis in tumor cells [40, 101]. AMF also promotes tumor progression *in vivo* through the paracrine method: it regulates proliferation of fibroblast, induces angiogenesis, and potentiates the physiologic action of VEGF [48, 81, 102, 103]. On endothelial cell AMF promotes cell locomotion, up regulating the expression of VEGFR, and the formation of tube-like structures mimicking angiogenesis in three-dimensional collagen gels [48].

Autocrine-produced factors regarding metastasis stimulate tumor cell proliferation, migration, adhesion, or secretion of proteolytic enzymes in a direct way, whereas paracrine factors contribute to the tumor development indirectly, affecting the microenvironment of the tumor. Tumor metastasis is the effect of a network of multiple factors, and these interactions are much more complicated *in vivo*, which must be taken into account when thinking of therapeutic strategies.

2.8 Conclusion

Tumor progression and metastasis are complicated biological phenomena which include many kinds of tumor- and host-derived factors. There is now considerable evidence to indicate that cancers and the microenvironment express the components of the AMF/AMFR axis and that this axis may have an important autocrine/paracrine functional role in regulating cancer development. In addition, while secreted AMF has not hitherto been detected in the secretion from normal cell lines, many clinical studies demonstrated that enzymatic activity of AMF is found in the serum and urine of patients who had malignant tumors including colorectal, breast, lung, kidney and gastrointestinal carcinomas [37–39]. This provides a basis for additional studies to evaluate the potential of the AMF axis as diagnostic tumor markers, and/or the exogenous AMF as a new target molecule for the cancer therapy, that is, neutralization of exogenous AMF would be helpful for various malignant tumor therapies.

References

1. Fidler IJ, Poste G: The cellular heterogeneity of malignant neoplasms: implications for adjuvant chemotherapy. Semin Oncol 12: 207–221, 1985
2. Schnipper L: Clinical implications of tumor-cell heterogeneity. N Engl J Med 314: 1423–1431, 1986

3. Liotta LA, Thorgeirsson UP, Garbisa S: Role of collagenases in tumor cell invasion. Cancer Metastasis Rev 1: 277–288, 1982
4. Strauli P, Haemmerli G: The role of cancer cell motility in invasion. Cancer Metastasis Rev 3: 127–141, 1984
5. Woolley DE: Collagenolytic mechanisms in tumor cell invasion. Cancer Metastasis Rev 3: 361–372, 1984
6. Erickson CA: Cell migration in the embryo and adult organism. Curr Opin Cell Biol 2: 67–74, 1990
7. Lauffenburger DA, Horwitz AF: Cell migration: a physically integrated molecular process. Cell 84: 359–369, 1996
8. Carr I: Lymphatic metastasis. Cancer Metastasis Rev 2: 307–317, 1983
9. Carr J, Dreher B, Carr I: Lymphatic metastasis; lymphangiochemotherapy of mammary cancer: ascitic form of rat mammary adenocarcinoma 13762. Clin Exp Metastasis 1: 29–38, 1983
10. Bouwens L, Jacobs R, Remels L, Wisse E: Natural cytotoxicity of rat hepatic natural killer cells and macrophages against a syngeneic colon adenocarcinoma. Cancer Immunol Immunother 27: 137–141, 1988
11. Weiss L, Orr FW, Honn KV: Interactions between cancer cells and the microvasculature: a rate-regulator for metastasis. Clin Exp Metastasis 7: 127–167, 1989
12. Crissman JD, Hatfield J, Schaldenbrand M, Sloane BF, Honn KV: Arrest and extravasation of B16 amelanotic melanoma in murine lungs: a light and electron microscopic study. Lab Invest 53: 470–478, 1985
13. Folkman J: Angiogenesis in cancer, vascular, rheumatoid and other disease. Nat Med 1: 27–31, 1995
14. Weidner N: Intratumor microvessel density as a prognostic factor in cancer. Am J Pathol 147: 9–19, 1995
15. Stoker M, Gherardi E: Regulation of cell movement: the motogenic cytokines. Biochim Biophys Acta 1072: 81–102, 1991
16. Lazar-Molnar E, Hegyesi H, Toth S, Falus A: Autocrine and paracrine regulation by cytokines and growth factors in melanoma. Cytokine 12: 547–554, 2000
17. Baggetto LG: Biochemical, genetic, and metabolic adaptations of tumor cells that express the typical multidrug-resistance phenotype: reversion by new therapies. J Bioenerg Biomembr 29: 401–413, 1997
18. Kerbel RS: Growth dominance of the metastatic cancer cell: cellular and molecular aspects. Adv Cancer Res 55: 87–132, 1990
19. Singh RK, Bucana CD, Gutman M, Fan D, Wilson MR, Fidler IJ: Organ site-dependent expression of basic fibroblast growth factor in human renal cell carcinoma cells. Am J Pathol 145: 365–374, 1994
20. Liotta LA, Kohn EC: The microenvironment of the tumour-host interface. Nature 411: 375–379, 2001
21. Fidler IJ: The organ microenvironment and cancer metastasis. Differentiation 70: 498–505, 2002
22. Paget S: The distribution of secondary growths in cancer of the breast. Lancet 1: 571–573, 1889
23. Fidler IJ, Ellis LM: Neoplastic angiogenesis–not all blood vessels are created equal. N Engl J Med 351: 215–216, 2004
24. Liotta LA, Tryggvason K, Garbisa S, Hart I, Foltz CM, Shafie S: Metastatic potential correlates with enzymatic degradation of basement membrane collagen. Nature 284: 67–68, 1980
25. Nakajima M, Morikawa K, Fabra A, Bucana CD, Fidler IJ: Influence of organ environment on extracellular matrix degradative activity and metastasis of human colon carcinoma cells. J Natl Cancer Inst 82: 1890–1898, 1990
26. Gutman M, Singh RK, Xie K, Bucana CD, Fidler IJ: Regulation of interleukin-8 expression in human melanoma cells by the organ environment. Cancer Res 55: 2470–2475, 1995

27. Aaronson SA: Growth factors and cancer. Science 254: 1146–1153, 1991
28. McCormick BA, Zetter BR: Interactions in angiogenesis and metastasis. Pharmacol Ther 53: 239–260, 1992
29. Rodeck U, Melber K, Kath R, Menssen HD, Varello M, Atkinson B, Herlyn M: Constitutive expression of multiple growth factor genes by melanoma cells but not normal melanocytes. J Invest Dermatol 97: 20–26, 1991
30. Nicolson GL: Cancer progression and growth: relationship of paracrine and autocrine growth mechanisms to organ preference of metastasis. Exp Cell Res 204: 171–180, 1993
31. Radinsky R: Paracrine growth regulation of human colon carcinoma organ-specific metastasis. Cancer Metastasis Rev 12: 345–361, 1993
32. Wilson J, Balkwill F: The role of cytokines in the epithelial cancer microenvironment. Semin Cancer Biol 12: 113–120, 2002
33. Liotta LA, Mandler R, Murano G, Katz DA, Gordon RK, Chiang PK, Schiffmann E: Tumor cell autocrine motility factor. Proc Natl Acad Sci USA 83: 3302–3306, 1986
34. Liotta LA, Rao CN, Barsky SH: Tumor invasion and the extracellular matrix. Lab Invest 49: 636–649, 1983
35. Turley EA: Molecular mechanisms of cell motility. Cancer Metastasis Rev 11: 1–3, 1992
36. Seiki M: The cell surface: the stage for matrix metalloproteinase regulation of migration. Curr Opin Cell Biol 14: 624–632, 2002
37. Gomm SA, Keevil BG, Thatcher N, Hasleton PS, Swindell RS: The value of tumour markers in lung cancer. Br J Cancer 58: 797–804, 1988
38. Baumann M, Kappl A, Lang T, Brand K, Siegfried W, Paterok E: The diagnostic validity of the serum tumor marker phosphohexose isomerase (PHI) in patients with gastrointestinal, kidney, and breast cancer. Cancer Invest 8: 351–356, 1990
39. Filella X, Molina R, Jo J, Mas E, Ballesta AM: Serum phosphohexose isomerase activities in patients with colorectal cancer. Tumour Biol 12: 360–367, 1991
40. Tsutsumi S, Hogan V, Nabi IR, Raz A: Overexpression of the autocrine motility factor/phosphoglucose isomerase induces transformation and survival of NIH-3T3 fibroblasts. Cancer Res 63: 242–249, 2003
41. Yanagawa T, Watanabe H, Takeuchi T, Fujimoto S, Kurihara H, Takagishi K: Overexpression of autocrine motility factor in metastatic tumor cells: possible association with augmented expression of KIF3A and GDI-ß. Lab Invest 84: 513–522, 2004
42. Watanabe H, Takehana K, Date M, Shinozaki T, Raz, A: Tumor cell autocrine motility factor is the neuroleukin/phosphohexose isomerase polypeptide. Cancer Res 56: 2960–2963, 1996
43. Kim JW, Dang CV: Multifaceted roles of glycolytic enzymes. Trends Biochem Sci 30: 142–150, 2005
44. Beutler E, West C, Britton HA, Harris J, Forman L: Glucosephosphate isomerase (GPI) deficiency mutations associated with hereditary nonspherocytic hemolytic anemia (HNSHA). Blood Cells Mol Dis 23: 402–409, 1997
45. Jeffery CJ, Bahnson BJ, Chien W, Ringe D, Petsko GA: Crystal structure of rabbit phosphoglucose isomerase, a glycolytic enzyme that moonlights as neuroleukin, autocrin motility factor, and differentiation mediator. Biochemistry 39: 955–964, 2000
46. Ravindranath Y, Paglia DE, Warrier I, Valentine W, Nakatani M, Brockway RA: Glucose phosphate isomerase deficiency as a cause of hydrops fetalis. N Engl J Med 316: 258–261, 1987
47. Matsumoto I, Staub A, Benoist C, Mathis D: Arthritis provoked by linked T and B cell recognition of a glycolytic enzyme. Science 286: 1732–1735, 1999
48. Funasaka T, Haga A, Raz A, Nagase H: Tumor autocrine motility factor is an angiogenic factor that stimulates endothelial cell motility. Biochem Biophys Res Commun 285: 118–128, 2001
49. Dang CV, Semenza GL: Oncogenic alterations of metabolism. Trends Biochem Sci 24: 68–72, 1999

50. Gurney ME, Apatoff BR, Spear GT, Baumel MJ, Antel JP, Bania MB, Reder AT: Neuroleukin: a lymphokine product of lectin-stimulated T cells. Science 234: 574–581, 1986
51. Xu W, Seiter K, Feldman E, Ahmed T, Chiao JW: The differentiation and maturation mediator for human myeloid leukemia cells shares homology with neuroleukin or phosphoglucose isomerase. Blood 87: 4502–4506, 1996
52. Yakirevich E, Naot Y: Cloning of a glucose phosphate isomerase/neuroleukin-like sperm antigen involved in sperm agglutination. Biol Reprod 62: 1016–1023, 2000
53. Cao MJ, Osatomi K, Matsuda R, Ohkubo M, Hara K, Ishihara T: Purification of a novel serine proteinase inhibitor from the skeletal muscle of white croaker (Argyrosomus argentatus). Biochem Biophys Res Commun 272: 485–489, 2000
54. Niinaka Y, Paku S, Haga A, Watanabe H, Raz A: Expression and secretion of neuroleukin/phosphohexose isomerase/maturation factor as autocrine motility factor by tumor cells. Cancer Res 58: 2667–2674, 1998
55. Silletti S, Watanabe H, Hogan V, Nabi IR, Raz A: Purification of B16-F1 melanoma autocrine motility factor and its receptor. Cancer Res 51: 3507–3511, 1991
56. Shimizu K, Tani M, Watanabe H, Nagamachi Y, Niinaka Y, Shiroishi T, Ohwada S, Raz A, Yokota J: The autocrine motility factor receptor gene encodes a novel type of seven transmembrane protein. FEBS Lett 456: 295–300, 1999
57. Nabi IR, Watanabe H, Raz A: Autocrine motility factor and its receptor: role in cell locomotion and metastasis. Cancer Metastasis Rev 11: 5–20, 1992
58. Silletti S, Raz A: Regulation of autocrine motility factor receptor expression in tumor cell locomotion and metastasis. Curr Top Microbiol Immunol 213: 137–169, 1996
59. Ohta Y, Minato H, Tanaka Y, Go T, Oda M, Watanabe Y: Autocrine motility factor receptor expression associates with tumor progression in thymoma. Int J Oncol 17: 259–264, 2000
60. Takanami I, Takeuchi K, Watanabe H, Yanagawa T, Takagishi K, Raz A: Significance of autocrine motility factor receptor gene expression as a prognostic factor in non-small-cell lung cancer. Int J Cancer 95: 384–387, 2001
61. Timar J, Raso E, Dome B, Ladanyi A, Banfalvi T, Gilde K, Raz A: Expression and function of the AMF receptor by human melanoma in experimental and clinical systems. Clin Exp Metastasis 19: 225–232, 2002
62. Folkman J, Klagsbrun M: Angiogenic factors. Science 235: 442–447, 1987
63. Risau W, Sariola H, Zerwes HG, Sasse J, Ekblom P, Kemler R, Doetschman T: Vasculogenesis and angiogenesis in embryonic-stem-cell-derived embryoid bodies. Development 102: 471–478, 1988
64. Folkman J, Shing Y: Angiogenesis. J Biol Chem 267: 10931–10934, 1992
65. Liekens S, De Clercq E, Neyts J: Angiogenesis: regulators and clinical applications. Biochem Pharmacol 61: 253–270, 2001
66. Zetter BR: Angiogenesis and tumor metastasis. Annu Rev Med 49: 407–424, 1998
67. Weidner N, Folkman J, Pozza F, Bevilacqua P, Allred EN, Moore DH, Meli S, Gasparini G: Tumor angiogenesis: a new significant and independent prognostic indicator in early-stage breast carcinoma. J Natl Cancer Inst 84: 1875–1887, 1992
68. Vermeulen PB, Gasparini G, Fox SB, Colpaert C, Marson LP, Gion M, Belien JA, de Waal RM, Van Marck E, Magnani E, Weidner N, Harris AL, Dirix LY: Second international consensus on the methodology and criteria of evaluation of angiogenesis quantification in solid human tumours. Eur J Cancer 38: 1564–1579, 2002
69. Shibuya M: Role of VEGF-flt receptor system in normal and tumor angiogenesis. Adv Cancer Res 67: 281–316, 1995
70. Ferrara N: VEGF and the quest for tumour angiogenesis factors. Nat Rev Cancer 2: 795–803, 2002
71. Hanahan D, Folkman J: Patterns and emerging mechanisms of the angiogenic switch during tumorigenesis. Cell 86: 353–364, 1996

72. Passaniti A, Taylor RM, Pili R, Guo Y, Long PV, Haney JA, Pauly RR, Grant DS, Martin GR: A simple, quantitative method for assessing angiogenesis and antiangiogenic agents using reconstituted basement membrane, heparin, and fibroblast growth factor. Lab Invest 67: 519–528, 1992
73. Senger DR, Galli SJ, Dvorak AM, Perruzzi CA, Harvey VS, Dvorak HF: Tumor cells secrete a vascular permeability factor that promotes accumulation of ascites fluid. Science 219: 983–985, 1983
74. Connolly DT: Vascular permeability factor: a unique regulator of blood vessel function. J Cell Biochem 47: 219–223, 1991
75. Nagy JA, Masse EM, Herzberg KT, Meyers MS, Yeo KT, Yeo TK, Sioussat TM, Dvorak HF: Pathogenesis of ascites tumor growth: vascular permeability factor, vascular hyperpermeability, and ascites fluid accumulation. Cancer Res 55: 360–368, 1995
76. de Vries C, Escobedo JA, Ueno H, Houck K, Ferrara N, Williams LT: The fms-like tyrosine kinase, a receptor for vascular endothelial growth factor. Science 255: 989–991, 1992
77. Millauer B, Wizigmann-Voos S, Schnurch H, Martinez R, Moller NP, Risau W, Ullrich A: High affinity VEGF binding and developmental expression suggest Flk-1 as a major regulator of vasculogenesis and angiogenesis. Cell 72: 835–846, 1993
78. Plate KH, Breier G, Weich HA, Risau W: Vascular endothelial growth factor is a potential tumour angiogenesis factor in human gliomas *in vivo*. Nature 359: 845–848, 1992
79. Hatva E, Kaipainen A, Mentula P, Jaaskelainen J, Paetau A, Haltia M, Alitalo K: Expression of endothelial cell-specific receptor tyrosine kinases and growth factors in human brain tumors. Am J Pathol 146: 368–378, 1995
80. Warren RS, Yuan H, Matli MR, Gillett NA, Ferrara N: Regulation by vascular endothelial growth factor of human colon cancer tumorigenesis in a mouse model of experimental liver metastasis. J Clin Invest 95: 1789–1797, 1995
81. Funasaka T, Haga A, Raz A, Nagase H: Autocrine motility factor secreted by tumor cells upregulates vascular endothelial growth factor receptor (Flt-1) expression in endothelial cells. Int J Cancer 101: 217–223, 2002
82. Kanno S, Oda N, Abe M, Terai Y, Ito M, Shitara K, Tabayashi K, Shibuya M, Sato Y: Roles of two VEGF receptors, Flt-1 and KDR, in the signal transduction of VEGF effects in human vascular endothelial cells. Oncogene 19: 2138–2146, 2000
83. Soker S, Kaefer M, Johnson M, Klagsbrun M, Atala A, Freeman MR: Vascular endothelial growth factor-mediated autocrine stimulation of prostate tumor cells coincides with progression to a malignant phenotype. Am J Pathol 159: 651–659, 2001
84. Barleon B, Siemeister G, Martiny-Baron G, Weindel K, Herzog C, Marme D: Vascular endothelial growth factor up-regulates its receptor fms-like tyrosine kinase 1 (FLT-1) and a soluble variant of FLT-1 in human vascular endothelial cells. Cancer Res 57: 5421–5425, 1997
85. Semenza GL: HIF-1 and human disease: one highly involved factor. Genes Dev 14: 1983–1991, 2000
86. Harris AL: Hypoxia – a key regulatory factor in tumour growth. Nat Rev Cancer 2: 38–47, 2002
87. Semenza GL: Targeting HIF-1 for cancer therapy. Nat Rev Cancer 3: 721–732, 2003
88. Wang GL, Jiang BH, Rue EA, Semenza GL: Hypoxia-inducible factor 1 is a basic-helix-loop-helix-PAS heterodimer regulated by cellular O2 tension. Proc Natl Acad Sci USA 92: 5510–5514, 1995
89. Huang LE, Gu J, Schau M, Bunn HF: Regulation of hypoxia-inducible factor 1alpha is mediated by an O2-dependent degradation domain via the ubiquitinproteasome pathway. Proc Natl Acad Sci USA 95: 7987–7992, 1998
90. Kallio PJ, Wilson WJ, O'Brien S, Makino Y, Poellinger L: Regulation of the hypoxia-inducible transcription factor 1alpha by the ubiquitin-proteasome pathway. J Biol Chem 274: 6519–6525, 1999
91. Maxwell PH, Wiesener MS, Chang GW, Clifford SC, Vaux EC, Cockman ME, Wykoff CC, Pugh CW, Maher ER, Ratcliffe PJ: The tumour suppressor protein VHL targets hypoxia-inducible factors for oxygen-dependent proteolysis. Nature 399: 271–275, 1999

92. Ivan M, Kondo K, Yang H, Kim W, Valiando J, Ohh M, Salic A, Asara JM, Lane WS, Kaelin WG Jr: HIF-1alpha targeted for VHL-mediated destruction by proline hydroxylation: implications for O2 sensing. Science 292: 464–468, 2001
93. Jaakkola P, Mole DR, Tian YM, Wilson MI, Gielbert J, Gaskell SJ, Kriegsheim Av, Hebestreit HF, Mukherji M, Schofiels CJ, Maxwell PH, Pugh CW, Ratcliffe PJ: Targeting of HIF-1alpha to the von Hippel-Lindau ubiquitylation complex by O2-regulated prolyl hydroxylation. Science 292: 468–472, 2001
94. Sivitz WI, Lund DD, Yorek B, Grover-McKay M, Schmid PG: Pretranslational regulation of two cardiac glucose transporters in rats exposed to hypobaric hypoxia. Am J Physiol 263: E562–E569, 1992
95. Semenza GL, Roth PH, Fang HM, Wang GL: Transcriptional regulation of genes encoding glycolytic enzymes by hypoxia-inducible factor 1. J Biol Chem 269: 23757–23763, 1994
96. Daly EB, Wind T, Jiang XM, Sun L, Hogg PJ: Secretion of phosphoglycerate kinase from tumour cells is controlled by oxygen-sensing hydroxylases. Biochim Biophys Acta 1691: 17–22, 2004
97. Yoon DY, Buchler P, Saarikoski ST, Hines OJ, Reber HA, Hankinson O: Identification of genes differentially induced by hypoxia in pancreatic cancer cells. Biochem Biophys Res Commun 288: 882–886, 2001
98. Niizeki H, Kobayashi M, Horiuchi I, Akakura N, Chen J, Wang J, Hamada JI, Seth P, Katoh H, Watanabe H, Raz A, Hosokawa M: Hypoxia enhances the expression of autocrine motility factor and the motility of human pancreatic cancer cells. Br J Cancer 86: 1914–1919, 2002
99. Krishnamachary B, Berg-Dixon S, Kelly B, Agani F, Feldser D, Ferreira G, Iyer N, LaRusch J, Pak B, Taghavi P, Semenza GL: Regulation of colon carcinoma cell invasion by hypoxia-inducible factor 1. Cancer Res 63: 1138–1143, 2003
100. Funasaka T, Yanagawa T, Hogan V, Raz A: Regulation of phosphoglucose isomerase/autocrine motility factor expression by hypoxia. FASEB J 19: 1422–1430, 2005
101. Haga A, Funasaka T, Niinaka Y, Raz A, Nagase H: Autocrine motility factor signaling induces tumor apoptotic resistance by regulations Apaf-1 and Caspase-9 apoptosome expression. Int J Cancer 107: 707–714, 2003
102. Silletti S, Raz A: Autocrine motility factor is a growth factor. Biochem Biophys Res Commun 194: 446–457, 1993
103. Tsutsumi S, Yanagawa T, Shimura T, Fukumori T, Hogan V, Kuwano H, Raz A: Regulation of cell proliferation by autocrine motility factor/phosphoglucose isomerase signaling. J Biol Chem 278: 32165–32172, 2003

Chapter 3
Targeting Signaling Pathways – In the Search of Melanoma's Achilles' Heel

Nikolas K. Haass, Christoph Hoeller and Meenhard Herlyn

Abstract Melanoma is the most aggressive form of skin cancer and is highly resistant to conventional chemotherapy, immunotherapy and targeted therapy. The prognosis for metastatic melanoma remains dismal with average survival rates of 6–10 months. Dacarbazine with response rates of less than 10% and a median progression free survival of 2 months is currently the only standard agent. Despite promising results with combination chemotherapy *in vitro*, and better response rates in patients, no randomized clinical trial has shown a survival advantage over single agent dacarbazine. Thus, new therapeutic targets are urgently needed to improve the dismal prognosis of this disease. The mitogen activated protein kinase (MAPK) pathway is constitutively active in most melanomas. The finding that over 60% of melanomas harbor the activating $BRAF^{V600E}$ mutation has raised expectations for the targeted therapy of melanoma. Small molecule signalling pathway inhibitors are now available for BRAF, $BRAF^{V600E}$, NRAS, MEK, mTOR, VEGF and others. In this review we discuss the role of targeting various constituents of the MAPK pathway and of mammalian target of rapamycin (mTOR). Finally we discuss the alternative approach of targeting melanoma stem cells as a putative therapy.

Keywords Targeted therapy · MAPKinase · BRAF · Activating mutation · c-KIT · MEK · PI3K · mTOR · Melanoma stem cells · Multidrug resistance

3.1 Introduction

3.1.1 Melanoma

Melanoma is a malignant tumor which derives from transformed melanocytes. Traditionally, the development of melanoma and its progression is described in

---

N.K. Haass
The Wistar Institute, 3601 Spruce Street, Philadelphia, PA 19104, USA, Centenary Institute of Cancer Medicine and Cell Biology, Royal Prince Alfred Hospital, University of Sydney, Newtown, NSW 2042, Australia
e-mail: N.Haass@centenary.org.au

M. Bar-Eli (ed.), *Regulation of Gene Expression in the Tumor Environment*,

six steps: (1) common acquired melanocytic nevus; (2) melanocytic nevus with lentiginous melanocytic hyperplasia, i.e., aberrant differentiation; (3) melanocytic nevus with aberrant differentiation and melanocytic nuclear atypia, i.e., melanocytic dysplasia; (4) the radial growth phase (RGP) of primary melanoma; (5) the vertical growth phase (VGP) of primary melanoma; and (7) metastatic melanoma [1]. Melanoma is mainly found in the skin, but in rare cases can also develop in the mucosa (of various localizations), conjunctiva and uvea, the inner ear, as well as in the meninges. Typically melanomas are pigmented, but there are also amelanotic melanomas which are often associated with delayed diagnosis and treatment. Approximately 69,000 new cases of melanoma plus an additional approximate 50,000 new cases of melanoma *in situ* were estimated for the United States in 2006 [2]. Thin melanomas below 1 mm of tumor thickness have a good prognosis and excision is curative in over 95% of all patients [2]. However, melanomas frequently metastasize early – i.e. at a thickness of as little as 1 mm and thus at a very low tumor mass – which leads to the dismal prognosis of melanoma. Patients with lymph node metastases or distant metastases have a relative five-year survival rate of 64% and 16% respectively [2] and a median survival of 12 months and 4–6 months respectively [3]. And while melanoma constitutes only 4% of skin cancers it is responsible for approximately 90% of skin cancer related deaths – approximately 10,700 people were estimated to die from melanoma in the United States in 2006 [2]. Importantly, many of these patients are of young age and although melanoma is the ninth most common cancer in the United States, it ranks second among solid tumors in terms of years of productive life lost.

3.1.2 Therapy of Metastatic Melanoma

Little progress has been made in the treatment of metastatic melanoma because of the absence of an effective systemic therapy. Where feasible, surgical resection of local recurrences, in-transit metastases (metastases within the range between the primary tumor and the regional lymph node station), and complete node dissection in the case of regional lymph node metastasis are recommended to lower the risk of developing further metastatic disease [3, 4]. An alternative for the treatment of non-resectable disease confined to a limb is hyperthermic isolated limb perfusion which is typically performed with melphalan (L-Phenylalanine Mustard) and more recently in combination with cytokines like tumor necrosis factor α (TNFα) or interferon α. Complete remissions are seen in 25–75% of patients but neither regimen has been demonstrated to improve overall survival [5]. Radiotherapy remains the treatment of choice for palliation of multiple cerebral metastases and for non-resectable bone metastases [4].

Currently, the only FDA-approved chemotherapeutic drug for the systemic therapy of metastatic melanoma is the alkylating agent dacarbazine (DTIC), which has a clinical response rate of less than 10% and hardly ever leads to durable complete remissions [6]. Efforts have been made to improve response and survival over those obtained with single-agent dacarbazine by the combination of several cytostatic or cytotoxic drugs. While some of these combinations did improve the overall response rate, they also showed a significantly higher number of adverse effects without any survival advantage over single agent DTIC [7].

The same is true for biochemotherapy, the combination of cytotoxic drugs with interleukin-2 and/or interferon-α. An improved response rate of up to 65% was associated with a significant increase in adverse events, but failed to demonstrate a survival advantage in the majority of studies performed [8]. Most importantly none of these regimens proved to be superior to high-dose bolus IL-2, so far the only cytokine based therapy that has received FDA approval for the treatment of metastatic melanoma. While the overall response rate in a phase II study was only 16%, a complete response was seen in 6% of patients and the median duration of response for these patients exceeds 59 months [8]. Unfortunately, the toxic events associated with this treatment highly limit the number of eligible patients.

There is a wealth of other immunologic approaches to melanoma treatment, including vaccination, blocking of cell surface receptors and treatment with cytokines. While some of these trials demonstrated an immunological response, the clinical response was confined to a small number of patients and demonstration of an impact on patient survival is yet missing [9].

A novel strategy to achieve improved results in the treatment of metastatic melanoma is the selection and alteration of specific molecules, responsible for uncontrolled growth, survival or other hallmarks of the malignant phenotype. One of the earliest studies in this field used an antisense oligonucleotide to target the antiapoptotic molecule Bcl-2 (oblimersen sodium, Genta Inc.). The randomized trial that compared treatment with DTIC to a combination of oblimersen sodium and DTIC included 771 patients and is the largest phase III trial so far performed in metastatic melanoma. The combination arm only showed a non-significant trend towards superiority in survival; however, subgroup analysis revealed a significant survival advantage in patients without elevated LDH. The secondary endpoints of overall response rate and response durability were significantly superior in the combination arm for the whole study population and all subgroups [10]. Further trials in different clinical settings will be necessary to reveal which patients will benefit most from this therapy.

3.2 Signaling Pathways and Their Inhibitors

Beyond the apoptosis pathway, several other highly interesting signaling cascades have been identified as candidates for 'targeted therapy' of metastatic melanoma over the last few years and are currently entering or undergoing clinical trials. In this review we discuss the role of targeting various constituents of the Mitogen Activated Protein Kinase Pathway (MAPK pathway) and of the phosphatidylinositol 3-kinase (PI3K)/Akt/mammalian target of rapamycin (mTOR) pathway.

3.2.1 The Mitogen Activated Protein Kinase Pathway (MAPK Pathway)

Cells respond to extracellular signals through the interaction of growth factors with their corresponding cell surface receptors by transmitting intracellular signals to

between cancer cells and their microenvironment, leading to cancer progression and metastasis.

4.1 ErbB2 Signaling and Its Role in Cancer

ErbB2 belongs to the epidermal growth factor receptor (EGFR) family. This family plays essential functions in multiple cellular processes including cell lineage determination, cellular proliferation, cellular survival, and organ morphogenesis [2, 3]. The EGFR family of receptor tyrosine kinases are comprised of EGFR (ErbB1), ErbB2 (HER2/neu), ErbB3 (HER3) and ErbB4 (HER4). ErbB2, has been implicated in a variety of human cancers. The c-erbB2 gene encodes a 185 kilodalton protein which is classified as an orphaned-receptor tyrosine kinase, since no ErbB2 ligand has been identified. Under normal conditions, extracellular ligands for EGFR, ErbB3, and ErbB4 will bind to their cognate receptors leading to receptor dimerization (either homo- or hetero-dimerization) with one another and result in receptor tyrosine trans-phosphorylation. Of the four members of the EGFR family, ErbB2 is the preferred binding partner of EGFR family receptors [4–6]. Receptor phosphorylations trigger their activation and lead to phosphorylation of intracellular tyrosine kinases [7–10], resulting in activation of downstream signaling cascades. Specifically, over-expression of ErbB2 can lead to deregulation of downstream pathways, such as the Ras/Raf/Extracellular signal Regulated Kinase (ERK) pathway (for proliferation), the phosphatidylinositol 3-kinase (PI3K)/Akt pathway (for survival and metabolism), and the Janus kinase/Signal Transducer and Activator of Transcription (Jak/STAT) pathway (for survival and cytokine responses) [11–14]. In addition to its classical function of activating downstream signaling pathways/networks, ErbB2 has recently been shown to translocate to the nucleus where it binds target promoters initiating transcription [15]. The complex nature of ErbB2 functions allows for a plethora of biological outcomes.

Pathologically, activating mutants and over-expression of EGFR family members contribute to cancer development by turning on complex signaling networks, driving numerous cancer promoting functions. Specifically, ErbB2 gene amplification and/or over-expression are seen in breast, ovarian, gastric, lung, liver, bladder, and several other types of cancers [16–21]. Among breast cancers, 20–30% over-express ErbB2 [16]. ErbB2 gene amplification and/or over-expression in breast cancer confers poor patient survival due to the detrimental effects, including, but not limited to, increased metastatic potential and therapeutic resistance [22–25]. Clinically, detection of *erbB2* gene amplification is a FDA approved procedure that is routinely conducted for breast cancer patients. ErbB2 expression profiling is performed in order to identify patients who may have a poor prognosis and to provide treatment options, such as Herceptin, a targeting antibody to the ErbB2 receptor [26–28]. To gain new insights on how to better control ErbB2 over-expressing breast cancers, it is important for us to understand how ErbB2 over-expression provokes breast cancer progression through both intracellular signaling and modulation of the extracellular environment.

4.2 The Impact of ErbB2 Over-Expression on Cancer Cells

The identification of ErbB2 over-expression in cancer has generated a wealth of information on the transformation process. ErbB2 over-expression contributes to essentially every step in the malignant process, thus focuses the importance of understanding the downstream pathways involved in each phase of cancer progression. The impact of ErbB2 signaling in cancer cells is highlighted here on three interwoven processes: increased cellular proliferation, acquisition of metastasis-related properties and resistance to apoptosis and therapeutic agents.

4.2.1 Proliferation

Malignant cells have altered properties that allow them to evade normal growth inhibitory signals and permit them to proliferate autonomously. ErbB2 over-expression facilitates this aberrant growth and further drives tumor progression (Fig. 4.1A).

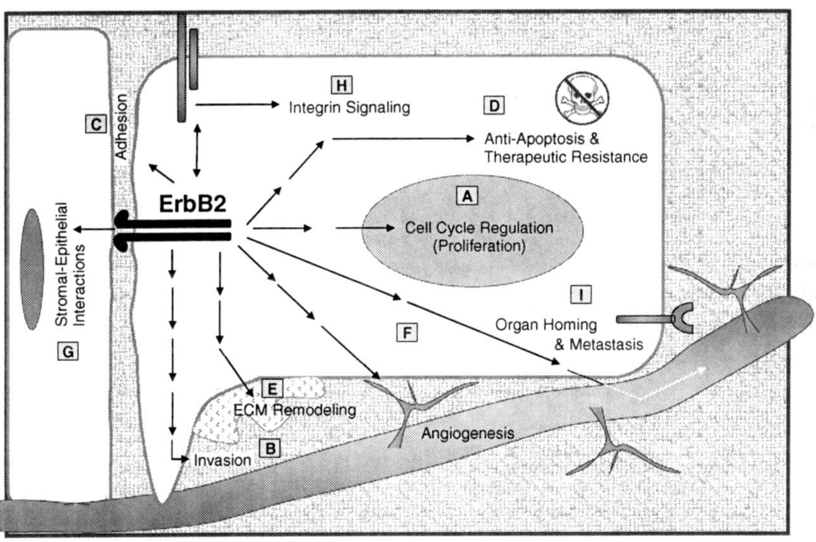

Fig. 4.1 Biology of multi-facetted ErbB2 signaling. ErbB2 signaling impacts many different cellular and microenvironment responses shown above. **A.** ErbB2 can signal through many cell cycle regulatory proteins promoting increased proliferation. **B.** Increased invasion of the ECM allows for ErbB2 over-expressing cells to enter circulation resulting in increased incidence of metastasis. **C.** ErbB2 can mediate cell adhesion, resulting in stromal detachment and increased endothelial attachment. **D.** ErbB2 signaling can result in anti-apoptosis and therapeutic resistance. **E.** Increased secretion of Matrix Metalloproteases by ErbB2 signaling leads to extracellular matrix degredation. **F.** ErbB2 can both transcriptionally and translationally upregulate VEGF production leading to increased angiogenesis. **G.** ErbB2 signaling can effect stromal-epithelial interactions that govern polarity and adhesion. **H.** Cooperation between ErbB2 and integrins promotes intracellular signaling based on extracellular conditions. **I.** ErbB2 mediated expression of proteins, such as CXCR4, can direct metastatic cell homing to organs such as bone, brain and lung

As previously mentioned, ErbB2 signaling contributes to oncogenesis through multiple signaling networks. Normal cell proliferation depends on a precisely regulated set of events that determine when a cell should grow and divide. Deviations in the signaling network controlling cell proliferation by over-expressed oncogenic ErbB2 can stimulate excessive cell division, leading to an increased tumor mass [29]. The role of ErbB2 in promoting proliferation and tumor growth has been clearly shown in *neu* (rat homologue of ErbB2) transgenic mice that develop tumors with almost 100% occurrence [30, 31]. Supporting its importance, multiple studies have shown that interference of ErbB2 or ErbB2 signaling activities results in anti-proliferation of the targeted cells [32–38].

Mechanistically, ErbB2 over-expression has downstream targets essential for the regulation of the cell cycle. These targets include c-Myc, cyclin D1, cell cycle inhibitors $p21^{cip1}$, $p27^{Kip1}$ [37, 39, 40], the cyclin E/CDK2 complex, and other regulators of cell proliferation [41]. An examination of ErbB2 transcriptional and translational responses observed increased MAPK signaling, increased CDK activity, down-regulation of CDK inhibitors and reduced cellular adhesion. These results attributed ErbB2 signaling to the promotion of tumorgenesis through anchorage-independent cellular proliferation [42].

4.3 Invasion and Metastasis

The formations of metastases are life-threatening events that are very common in the late stages of cancer. ErbB2 over-expression in breast cancers correlates with increased metastasis in patients [16] and is also shown through neu transgenic mice, which readily develop metastasis [30, 43, 44] (Fig. 4.1B). The spread of metastases can occur through either the circulatory and/or the lymphatic systems. In order for cancer cells to reach systemic circulation, they must first gain the ability to travel to and then enter the vessels. This process requires cells to migrate away from the primary site and invade the surrounding tissues. These processes often occur through the acquisition of metastatic properties such as increased heterologous cell adhesion, motility, migration, and invasion. ErbB2 has been shown to potentate all of these metastasis-related properties in multiple systems [31, 45]. For example, stimulation of breast cancer cells with EGF or Heregulin have shown an increase in cellular migration [46]. Conversely, ErbB2 receptor inactivation demolished cellular migration after EGF and Heregulin treatment, demonstrating that a functional ErbB2 is essential for migration.

Reduced homophilic cell–cell adhesion and increased heterophilic adhesion (e.g., cancer cell adhesion to endothelial cells) can promote metastasis [47] (Fig. 4.1C). For example, over-expression of ErbB2 leads to a decrease in E-cadherin, an adhesion molecule, through transcriptional downregulation [48]. Loss of E-cadherin then prevents cell–cell homophilic adhesion in mammary epithelial cells, and results in the loss of cell polarity and consequent transformation. The stromal components can also mediate adhesion of epithelial cells. Adhesion of epithelial cells to collagen IV, lamin, and fibronectin reduce the ability of EGF to bind to EGFR/ErbB2 heterodimers, however, adhesion to collagen I does not [49]. This suggests that specific adhesion signals or cell-matrix interactions differentially modulate ErbB2-mediated signaling. On the other hand, ErbB2 can also increase the adhesion of breast cancer cells to endothelial cells. Access to endothelial cells allows cancer cells to travel into or out of the blood stream, important steps in the metastasis process [31].

Once cancer cells become invasive, they are able to penetrate their basement membranes gaining access to the circulation systems. In order for invasion to occur the basement membrane must be degraded. Basement membrane degredation occurs primarily through the secretion of extracellular proteases, such as matrix metalloproteases (MMPs) and serine proteases (uPA), which cleave components of the extracellular matrix. ErbB2 over-expression increases secretion of these proteases, which will be further explored later on in this chapter.

ErbB2 downstream signaling is also involved in promoting invasion. For example, inhibitors to EGFR, ErbB2 and PI3K pathways decreased pancreatic cells' invasiveness and decreased their adhesion to collagen I [50]. The role of the PI3K in ErbB2-mediated invasion is indicated in a study that utilized an ErbB2 homodimer and ErbB2/EGFR heterodimer inducible system [51]. This study showed that ErbB2 homodimers signaled primarily through the Ras/MAPK pathway, while ErbB2 heterodimers signaled through the PI3K and PLCγ pathways. Both ErbB2 homo- and hetero-dimers induced disruption of acini-like formations, however, only cells with induced ErbB2/EGFR heterodimers were able to invade the surrounding matrix in a 3D cell culture system. These data suggest that the PI3K and PLCγ pathways contribute to the ErbB2-induced increase of cell invasion.

4.3.1 Apoptosis and Therapeutic Resistance

In addition to its proliferative capacities, ErbB2 over-expression results in resistance to apoptosis and increased cellular survival (Fig. 4.1D). This often results in therapeutic resistance to standard cancer treatments, such as chemotherapy, hormone therapy, radiation therapy, cytokine treatments, and even targeted therapies.

Clinical studies have demonstrated that over-expression of ErbB2 results in resistance to chemotherapeutic treatments including; cyclophosphomide-methotrexate-5′ fluorouracil (CMF), taxane therapies (Taxol, Taxotere), and epirubicin [23, 24, 52, 53]. These studies along with laboratory experiments have specified ErbB2 as a marker of drug resistance. This is further supported by studies showing that the treatment of breast cancer cells with an ErbB2 antagonist, Herceptin, allowed for re-sensitization of once resistant cells to Taxol chemotherapy [23, 24]. Moreover, ErbB2 over-expressing breast cancer xenografts in mice are sensitized to doxorubicin treatment following ErbB2 downregulation by Herceptin [54, 55].

Numerous studies identify the relationship between ErbB2 and hormone receptors [56–60]. Transfection of cell lines with ErbB2 renders cells resistance to hormonal therapies such as tamoxifen, demonstrating that ErbB2 oncogenic signaling leads to antiestrogen therapy resistance [61]. Also, breast cancer cells that overexpress ErbB2 by transfection showed MAPK pathway activation and consequent tamoxifen resistance. Tamoxifen sensitivity was restored following treatment with the EGFR/ErbB2 inhibitors, gefitinib (ZD1839) and erlotinib (OSI-774) [22].

ErbB2 over-expression also contributes to radiotherapy resistance. Applying antisense oligonucleotides to both ras and raf-1 in ErbB2 over-expressing cells restored their radiosensitivity, suggesting that ErbB2 downstream RAS/Raf/MAPK pathway is responsible for radio-resistance [62, 63]. Additionally, interruption of ErbB2-mediated NF-κB activation re-sensitized cells to gamma irradiation, inhibited proliferation and promoted apoptosis [64, 65].

Taken together, ErbB2 over-expression can activate multiple down-stream signaling events that promote cancer cell proliferation, increases invasion and metastasis potential, and confers resistance to apoptosis and to cancer therapeutics.

4.4 The Impact of ErbB2 Over-Expression on the Tumor Microenvironment

The area and components surrounding tumor cells set up a microenvironment paramount to tumor cell growth, survival, invasion and metastasis. The ability of a cancer cell to modulate this environment contributes to tumor progression. For a tumor to grow, expand, and metastasize, at least two key components in tumor microenvironment need to be altered. A tumor needs to degrade its surrounding extracellular matrix (ECM) and also increase blood supply to facilitate growth. ErbB2 over-expression has been implicated in both of these processes.

4.4.1 Degradation of ECM

Interactions between cells and their extracellular matrix are tightly regulated and drive a variety of biological outcomes. In cancer, tumor cells often gain the ability to degrade their extracellular matrix in order to expand in size and facilitate adhesion, migration and invasion.

Matrix Metalloproteases (MMPs) are a family of collagenases that directly remodel the extracellular matrix. These proteases are capable of degrading all components of the extracellular matrix, as well as, non-classical substrates such as growth factors, cytokines and cell adhesion molecules, such as E-cadherin [66, 67]. Over-expression of ErbB2 can upregulate the protease activities of both MMP-2 and MMP-9, leading to a more invasive phenotype [68]. Additionally, stimulation of breast cancer cell by heregulin (ligand for ErbB2/ErbB3 dimers), led to increased secretion of urokinase plasminogen activator (uPA) [69]. uPA selectively cleaves plasminogen and degrades the basement membrane. This observation is supported by clinical studies that correlate ErbB2 over-expression with uPA/PAI-1 in patient samples [70]. These studies provide examples to how ErbB2 over-expression can result in alterations in the extracellular matrix that surround a tumor cell (Fig. 4.1E). These alterations contribute to tissue remodeling and lead to increased cellular invasiveness.

4.4.2 Angiogenesis

Breast cancer is typified as being exceedingly vascularized [71]. Increased vascularization occurs by increasing the formation of new blood vessels, or angiogenesis. Angiogenesis is a normal process tightly regulated within the body. Under pathological conditions, however, the balance between pro-angiogenic and anti-angiogenic factors is disturbed. Vascular Endothelial Growth Factor (VEGF) is the most potent pro-angiogenic factor known to date. ErbB2 over-expression mediates increased VEGF secretion and its impact to tumors and their extracellular environment is detrimental (Fig. 4.1F). As a tumor cell secretes VEGF, vascular networks are constructed and directed to support and feed the tumor mass. Newly formed tumor vasculature brings increased nutrients and oxygen, further promoting tumor growth and increased cellular proliferation. In turn, this increased tumor mass produces more VEGF, exponentially driving tumor progression. In addition to increased tumor growth, increased vascularization allows for systemic access of tumor cells, thus promoting metastasis. ErbB2 positive tumors are more vascularized than tumors expressing normal levels of ErbB2 and expectedly express higher levels of VEGF [72, 73]. ErbB2 in breast cancers also correlated with increased expression of the lymph-angiogenic factors, VEGF-C and VEGF-D [74]. Increased secretion of VEGF by ErbB2 could occur through both increased transcription and translation [75, 76]. Transcriptionally, increased secretion of VEGF occurs through ErbB2 transactivational upregulation of transcription factor Sp1, mediating VEGF transcriptional upregulation in NIH3T3 cells [75]. Translationally, ErbB2 activates mammalian target of rapamycin and phospho-p70S6 kinase (mTOR/p70S6K) through both the Erk/MAPK and PI3K pathways, leading to increased translation of VEGF mRNA in human breast cancer cells [76]. Pathways involved in VEGF upregulation provide opportunities for therapeutic intervention and ErbB2 targeted therapy. For example, using an ErbB2 targeting agent, Herceptin, together with Taxol, angiogenesis was more effectively inhibited [77]. Furthermore, another study showed that AEE788, a dual EGFR/ErbB2 kinase inhibitor, had both anti-tumor and anti-angiogenic activity [78].

4.5 Synergistic Interactions Between ErbB2 Over-Expressing Cancer Cells and Altered Tumor Microenvironment

The delicate balance between the extracellular environment and intracellular signaling provide homeostasis for normal growth and development. Signaling by both the extracellular environment and cancer cells can synergistically promote tumor progression and metastasis.

4.5.1 Stromal-Epithelial Interactions

Biologically, cells need to communicate effectively with their outside environment. The extracellular components dictate whether a cell should grow, divide, or even die. Stromal elements communicate with cell surface proteins creating a linked community. Specific genes are responsible for this network and are often involved in cancer. A cDNA microarray analysis study observed that ErbB2 over-expressing cell lines and primary breast cancer samples had differential expression in genes that regulate cell-matrix interactions (Fig. 4.1G). These genes included fibronectin, p-cadherin, falectin 1 & 3, and proline 4-hydroxylase [79].

Integrins are membrane bound proteins that allow for the attachment of cells to their extracellular matrix and to other cells. In addition, integrins connect signal transduction information from the extracellular matrix to the cell. Recent studies have demonstrated the dependence between ErbB2 and integrin signaling (Fig. 4.1H). β4 Integrin is required for ErbB2 signaling, and blockade of β4 prevented ErbB2-mediated tumorigenesis [80, 81]. Proteins responsible for integrin signaling provide a connection between extracellular information and intracellular signaling. Integrin linked kinase (ILK) controls extracellular matrix interactions and cell proliferation which are essential for normal skin development [82, 83]. A transgenic mouse model that over-expressed ErbB2 in the epidermis of the mice showed regulation of ILK expression by ErbB2. The dependence of ILK expression on ErbB2 connects the extracellular signaling pathway with the growth and proliferation pathway [84]. These mice displayed hyperplastic skin phenotypes and died shortly after birth. Additionally, ErbB2 over-expression correlated with ILK over-expression and dermal/hyperplastic localization.

Anomalies in the physical interactions between the stroma and epithelial cells can drive disease progression. For example, stromal HGF secretion can disrupt ErbB2 signaling in epithelial cells [85]. The disruption resulted in the degredation of cell–cell junctions and increased invasion, thereby enhancing the malignant phenotype. The invasion was abrogated by using an inhibitor to the MAPK pathway allowing for E-cadherin re-expression. This study demonstrated how epithelial cells that over-express ErbB2, along with stromal cells that secrete HGF can synergize to enhance the malignant phenotype.

In addition, the spatial dislocation of ErbB2 can result in the loss of contact between stromal and epithelial cells. Normally, ErbB2 is located to the basolateral surface of polarized epithelial cells where it interacts with the stroma. However, when Muc4, a membrane bound glycoprotein, is over-expressed, Muc4 will bind ErbB2 in its transmembrane region, sequestering it to the apical surface of the cell. Over-expression of Muc4 is known to dissociate cell–cell and cell-matrix interactions, and these finding provide a possible ErbB2-mediated mechanism [86]. Additionally, over-expression of Muc4 has resulted in the relocalization of E-cadherin from the lateral membrane to the apical membrane triggering loss of cell–cell contact and increased invasiveness [87].

To give another example of ErbB2-mediated cellular interactions, CD40 Knock-out mice that also over-express ErbB2 within the mammary gland (CD40 KO/ MMTV-NeuT), showed delayed tumor onset, when compared to mice that only over-expressed ErbB2 in the mammary gland (MMTV-NeuT) [88]. The delayed tumor onset is a consequence of impaired angiogenesis. Thus, CD40 expression in endothelial cells is required for proper vessel formation in the context of ErbB2 over-expressing mouse models. These compelling studies highlight how ErbB2 over-expression drives tumor progression through cooperation with multiple microenvironment alterations and explains the perilous outcomes that arise from ErbB2 over-expressing cancers.

4.5.2 Homing of ErbB2 Over-Expressing Cancer Cells via Upregulation of CXCR4

CXCR4 is a G-protein coupled receptor which has been implicated in metastasis. Its ligand, SDF-1α, is expressed in organs such as lymph nodes, liver, lung and bone marrow; all common sites of breast cancer metastasis [89, 90]. CXCR4 expression is believed to recruit cancer cells to areas/organs containing SDF-1α. This idea was supported by data showing that the migration of breast cancer cells was reduced following CXCR4 blockade in human bone marrow [91]. Co-expression and cross talk of CXCR4 and ErbB2 was reported in metastatic breast tumors and led to reduced overall survival (Fig. 4.1I). CXCR4 expression was constitutive due to ErbB2 inhibition of CXCR4 degredation [92]. Furthermore, this study demonstrated that when CXCR4 level was decreased, breast cancer cells had a reduced ability to metastasize to the lung in mouse model studies. Taken together this indicated that CXCR4 was required for SDF-1α mediated cell adhesion, invasion and metastasis. These observations culminated when it was demonstrated that CXCR4/ErbB2 expression in breast cancer cells increased the likelihood of metastasis to the bone marrow [93]. *In vivo* models showed that intracellular signaling of ErbB2 can lead to increased stability of CXCR4, and therefore allow increased adhesion, migration and invasion, as well as, metastatic homing to target organs. Together, it is clear that ErbB2 over-expression has the ability to directly affect intracellular signaling that drives extracellular responses.

4.6 Summary and Prospective

In summary, multitudes of studies have all demonstrated how ErbB2 over-expression in cancer cells can activate a complex cancer cell signaling network. This network impacts not only cancer cell properties but also impacts on the surrounding microenvironment of the cancer cell, thus contributing to various cancer phenotypes. Furthermore, ErbB2 over-expression can promote synergistic interplay between the cell–cell, cell-stromal, and cell-matrix, resulting in cancer progression and metastasis. Therefore, when we consider strategies to control ErbB2 over-expressing breast cancers, it is imperative to use anti-cancer therapies that effectively target both the cancer cells and their microenvironment, or target key nodes that impact on both. Recently, Herceptin treatment has been shown to directly affect stromal cells in addition to the originally targeted cancer cells [94], providing an excellent example of a successful anti-cancer therapy that targets both the cancer cells and their microenvironment. It is foreseeable that as we begin to have a better understanding on how the interplay between cancer cells and their microenvironment impact tumor formation and tumor progression, more powerful anti-cancer strategies and therapeutics will be developed that effectively target both.

Acknowledgments Although we aimed to encompass a large variety of published data, this review, inevitably, may have unintentionally omitted some important published articles. Here, we sincerely apologize for any omission and acknowledge the contributions of the entire research community. Dr. Yu is the Nylene Eckles Distinguished Professor in breast cancer research at the University of Texas M.D. Anderson Cancer Center.

References

1. Muthuswamy, S.K., et al., *ErbB2, but not ErbB1, reinitiates proliferation and induces luminal repopulation in epithelial acini.* Nat Cell Biol, 2001. **3**(9): 785–92.
2. Yarden, Y., *The EGFR family and its ligands in human cancer: signalling mechanisms and therapeutic opportunities.* Eur J Cancer, 2001. **37(Suppl 4)**: S3–8.
3. Negro, A., B.K. Brar, and K.F. Lee, *Essential roles of Her2/erbB2 in cardiac development and function.* Recent Prog Horm Res, 2004. **59**: 1–12.
4. Di Fiore, P.P., et al., *erbB-2 is a potent oncogene when overexpressed in NIH/3T3 cells.* Science, 1987. **237**(4811): 178–82.
5. Olayioye, M.A., et al., *ErbB-1 and ErbB-2 acquire distinct signaling properties dependent upon their dimerization partner.* Mol Cell Biol, 1998. **18**(9): 5042–51.
6. Riese, D.J., II, et al., *The cellular response to neuregulins is governed by complex interactions of the erbB receptor family.* Mol Cell Biol, 1995. **15**(10): 5770–76.
7. Basu, T., P.H. Warne, and J. Downward, *Role of Shc in the activation of Ras in response to epidermal growth factor and nerve growth factor.* Oncogene, 1994. **9**(12): 3483–91.
8. Gale, N.W., et al., *Grb2 mediates the EGF-dependent activation of guanine nucleotide exchange on Ras.* Nature, 1993. **363**(6424): 88–92.
9. Li, N., et al., *Guanine-nucleotide-releasing factor hSos1 binds to Grb2 and links receptor tyrosine kinases to Ras signalling.* Nature, 1993. **363**(6424): 85–88.
10. Wilkinson, M.G. and J.B. Millar, *Control of the eukaryotic cell cycle by MAP kinase signaling pathways.* Faseb J, 2000. **14**(14): 2147–57.
11. Cobb, M.H. and E.J. Goldsmith, *How MAP kinases are regulated.* J Biol Chem, 1995. **270**(25): 14843–46.
12. Fruman, D.A., R.E. Meyers, and L.C. Cantley, *Phosphoinositide kinases.* Annu Rev Biochem, 1998. **67**: 481–507.
13. Kisseleva, T., et al., *Signaling through the JAK/STAT pathway, recent advances and future challenges.* Gene, 2002. **285**(1–2): 1–24.
14. Aaronson, D.S. and C.M. Horvath, *A road map for those who don't know JAK-STAT.* Science, 2002. **296**(5573): 1653–55.
15. Wang, S.C., et al., *Binding at and transactivation of the COX-2 promoter by nuclear tyrosine kinase receptor ErbB-2.* Cancer Cell, 2004. **6**(3): 251–61.
16. Slamon, D.J., et al., *Studies of the HER-2/neu proto-oncogene in human breast and ovarian cancer.* Science, 1989. **244**(4905): 707–12.
17. Lacroix, H., et al., *Overexpression of erbB-2 or EGF receptor proteins present in early stage mammary carcinoma is detected simultaneously in matched primary tumors and regional metastases.* Oncogene, 1989. **4**(2): 145–51.

18. Voravud, N., et al., *Oncogene expression in cholangiocarcinoma and in normal hepatic development.* Hum Pathol, 1989. **20**(12): 1163–68.
19. Nigawara, K., et al., *[Expression of c-H-ras, c-erb B1 and c-erb B2 gene products in human bladder cancer].* Nippon Hinyokika Gakkai Zasshi, 1992. **83**(8): 1212–19.
20. Heinmoller, P., et al., *HER2 status in non-small cell lung cancer: results from patient screening for enrollment to a phase II study of herceptin.* Clin Cancer Res, 2003. **9**(14): 5238–43.
21. Falck, V.G. and W.J. Gullick, *c-erbB-2 oncogene product staining in gastric adenocarcinoma: an immunohistochemical study.* J Pathol, 1989. **159**(2): 107–11.
22. Shin, I., T. Miller, and C.L. Arteaga, *ErbB receptor signaling and therapeutic resistance to aromatase inhibitors.* Clin Cancer Res, 2006. **12**(3 Pt 2): 1008s–1012s.
23. Yu, D., et al., *Overexpression of ErbB2 blocks Taxol-induced apoptosis by upregulation of p21Cip1, which inhibits p34Cdc2 kinase.* Mol Cell, 1998. **2**(5): 581–91.
24. Yu, D., et al., *Overexpression of both p185c-erbB2 and p170mdr-1 renders breast cancer cells highly resistant to taxol.* Oncogene, 1998. **16**(16): 2087–94.
25. Yu, D., et al., *Overexpression of c-erbB-2/neu in breast cancer cells confers increased resistance to Taxol via mdr-1-independent mechanisms.* Oncogene, 1996. **13**(6): 1359–65.
26. Piccart-Gebhart, M.J., et al., *Trastuzumab after adjuvant chemotherapy in HER2-positive breast cancer.* N Engl J Med, 2005. **353**(16): 1659–72.
27. Robert, N., et al., *Randomized phase III study of trastuzumab, paclitaxel, and carboplatin compared with trastuzumab and paclitaxel in women with HER-2-overexpressing metastatic breast cancer.* J Clin Oncol, 2006. **24**(18): 2786–92.
28. Romond, E.H., et al., *Trastuzumab plus adjuvant chemotherapy for operable HER2-positive breast cancer.* N Engl J Med, 2005. **353**(16): 1673–84.
29. Pavelic, Z.P., et al., *c-myc, c-erbB-2, and Ki-67 expression in normal breast tissue and in invasive and noninvasive breast carcinoma.* Cancer Res, 1992. **52**(9): 2597–602.
30. Guy, C.T., et al., *Expression of the neu protooncogene in the mammary epithelium of transgenic mice induces metastatic disease.* Proc Natl Acad Sci USA, 1992. **89**(22): 10578–82.
31. Yu, D., et al., *Mechanisms of c-erbB2/neu oncogene-induced metastasis and repression of metastatic properties by adenovirus 5 E1A gene products.* Oncogene, 1992. **7**(11): 2263–70.
32. Lane, H.A., et al., *ErbB2 potentiates breast tumor proliferation through modulation of p27(Kip1)-Cdk2 complex formation: receptor overexpression does not determine growth dependency.* Mol Cell Biol, 2000. **20**(9): 3210–23.
33. Basso, A.D., et al., *Ansamycin antibiotics inhibit Akt activation and cyclin D expression in breast cancer cells that overexpress HER2.* Oncogene, 2002. **21**(8): 1159–66.
34. Moasser, M.M., et al., *The tyrosine kinase inhibitor ZD1839 ("Iressa") inhibits HER2-driven signaling and suppresses the growth of HER2-overexpressing tumor cells.* Cancer Res, 2001. **61**(19): 7184–88.
35. Motoyama, A.B., N.E. Hynes, and H.A. Lane, *The efficacy of ErbB receptor-targeted anticancer therapeutics is influenced by the availability of epidermal growth factor-related peptides.* Cancer Res, 2002. **62**(11): 3151–58.
36. Munster, P.N., et al., *Degradation of HER2 by ansamycins induces growth arrest and apoptosis in cells with HER2 overexpression via a HER3, phosphatidylinositol 3′-kinase-AKT-dependent pathway.* Cancer Res, 2002. **62**(11): 3132–37.
37. Neve, R.M., et al., *Effects of oncogenic ErbB2 on G1 cell cycle regulators in breast tumour cells.* Oncogene, 2000. **19**(13): 1647–56.
38. Yakes, F.M., et al., *Herceptin-induced inhibition of phosphatidylinositol-3 kinase and Akt Is required for antibody-mediated effects on p27, cyclin D1, and antitumor action.* Cancer Res, 2002. **62**(14): 4132–41.
39. Lenferink, A.E., et al., *ErbB2/neu kinase modulates cellular p27(Kip1) and cyclin D1 through multiple signaling pathways.* Cancer Res, 2001. **61**(17): 6583–91.
40. Lee, R.J., et al., *Cyclin D1 is required for transformation by activated Neu and is induced through an E2F-dependent signaling pathway.* Mol Cell Biol, 2000. **20**(2): 672–83.

41. Keyomarsi, K., et al., *Cyclin E and survival in patients with breast cancer.* N Engl J Med, 2002. **347**(20): 1566–75.
42. White, S.L., et al., *Cellular responses to ErbB-2 overexpression in human mammary luminal epithelial cells: comparison of mRNA and protein expression.* Br J Cancer, 2004. **90**(1): 173–81.
43. Moody, S.E., et al., *Conditional activation of Neu in the mammary epithelium of transgenic mice results in reversible pulmonary metastasis.* Cancer Cell, 2002. **2**(6): 451–61.
44. Muller, W.J., et al., *Single-step induction of mammary adenocarcinoma in transgenic mice bearing the activated c-neu oncogene.* Cell, 1988. **54**(1): 105–15.
45. Ritch, P.A., S.L. Carroll, and H. Sontheimer, *Neuregulin-1 enhances motility and migration of human astrocytic glioma cells.* J Biol Chem, 2003. **278**(23): 20971–78.
46. Wolf-Yadlin, A., et al., *Effects of HER2 overexpression on cell signaling networks governing proliferation and migration.* Mol Syst Biol, 2006. **2**: 54.
47. Jothy, S., et al., *Adhesion or anti-adhesion in cancer: what matters more?* Cancer Metastasis Rev, 1995. **14**(4): 363–76.
48. D'Souza, B. and J. Taylor-Papadimitriou, *Overexpression of ERBB2 in human mammary epithelial cells signals inhibition of transcription of the E-cadherin gene.* Proc Natl Acad Sci USA, 1994. **91**(15): 7202–06.
49. Woodward, T.L., et al., *Proliferation of mouse mammary epithelial cells in vitro: interactions among epidermal growth factor, insulin-like growth factor I, ovarian hormones, and extracellular matrix proteins.* Endocrinology, 2000. **141**(10): 3578–86.
50. Shirk, A.J. and R. Kuver, *Epidermal growth factor mediates detachment from and invasion through collagen I and Matrigel in Capan-1 pancreatic cancer cells.* BMC Gastroenterol, 2005. **5**: 12.
51. Zhan, L., B. Xiang, and S.K. Muthuswamy, *Controlled activation of ErbB1/ErbB2 heterodimers promote invasion of three-dimensional organized epithelia in an ErbB1-dependent manner: implications for progression of ErbB2-overexpressing tumors.* Cancer Res, 2006. **66**(10): 5201–08.
52. Gusterson, B.A., et al., *Prognostic importance of c-erbB-2 expression in breast cancer. International (Ludwig) Breast Cancer Study Group.* J Clin Oncol, 1992. **10**(7): 1049–56.
53. Jarvinen, T.A., et al., *Predictive value of topoisomerase IIalpha and other prognostic factors for epirubicin chemotherapy in advanced breast cancer.* Br J Cancer, 1998. **77**(12): 2267–73.
54. Albanell, J., F. Rojo, and J. Baselga, *Pharmacodynamic studies with the epidermal growth factor receptor tyrosine kinase inhibitor ZD1839.* Semin Oncol, 2001. **28**(5 Suppl 16): 56–66.
55. Pegram, M.D., et al., *Phase II study of receptor-enhanced chemosensitivity using recombinant humanized anti-p185HER2/neu monoclonal antibody plus cisplatin in patients with HER2/neu-overexpressing metastatic breast cancer refractory to chemotherapy treatment.* J Clin Oncol, 1998. **16**(8): 2659–71.
56. Saal, L.H., et al., *PIK3CA mutations correlate with hormone receptors, node metastasis, and ERBB2, and are mutually exclusive with PTEN loss in human breast carcinoma.* Cancer Res, 2005. **65**(7): 2554–59.
57. Konecny, G., et al., *Quantitative association between HER-2/neu and steroid hormone receptors in hormone receptor-positive primary breast cancer.* J Natl Cancer Inst, 2003. **95**(2): 142–53.
58. Stal, O., et al., *ErbB2 status and the benefit from two or five years of adjuvant tamoxifen in postmenopausal early stage breast cancer.* Ann Oncol, 2000. **11**(12): 1545–50.
59. Grunt, T.W., et al., *Bidirectional interactions between the estrogen receptor and the cerbB-2 signaling pathways: heregulin inhibits estrogenic effects in breast cancer cells.* Int J Cancer, 1995. **63**(4): 560–67.
60. Yang, Z., C.J. Barnes, and R. Kumar, *Human epidermal growth factor receptor 2 status modulates subcellular localization of and interaction with estrogen receptor alpha in breast cancer cells.* Clin Cancer Res, 2004. **10**(11): 3621–28.

61. Benz, C.C., et al., *Estrogen-dependent, tamoxifen-resistant tumorigenic growth of MCF-7 cells transfected with HER2/neu.* Breast Cancer Res Treat, 1992. **24**(2): 85–95.
62. Burke, H.B., et al., *Predicting response to adjuvant and radiation therapy in patients with early stage breast carcinoma.* Cancer, 1998. **82**(5): 874–77.
63. Pirollo, K.F., et al., *p53 mediated sensitization of squamous cell carcinoma of the head and neck to radiotherapy.* Oncogene, 1997. **14**(14): 1735–46.
64. Shao, R., et al., *Inhibition of nuclear factor-kappaB activity is involved in E1A-mediated sensitization of radiation-induced apoptosis.* J Biol Chem, 1997. **272**(52): 32739–42.
65. Zhou, B.P., et al., *HER-2/neu blocks tumor necrosis factor-induced apoptosis via the Akt/NF-kappaB pathway.* J Biol Chem, 2000. **275**(11): 8027–31.
66. Gearing, A.J., et al., *Processing of tumour necrosis factor-alpha precursor by metalloproteinases.* Nature, 1994. **370**(6490): 555–57.
67. Kajita, M., et al., *Membrane-type 1 matrix metalloproteinase cleaves CD44 and promotes cell migration.* J Cell Biol, 2001. **153**(5): 893–904.
68. Tan, M., J. Yao, and D. Yu, *Overexpression of the c-erbB-2 gene enhanced intrinsic metastasis potential in human breast cancer cells without increasing their transformation abilities.* Cancer Res, 1997. **57**(6): 1199–205.
69. Mazumdar, A., et al., *Heregulin regulation of urokinase plasminogen activator and its receptor: human breast epithelial cell invasion.* Cancer Res, 2001. **61**(1): 400–05.
70. Ryan, B.M., et al., *Survivin expression in breast cancer predicts clinical outcome and is associated with HER2, VEGF, urokinase plasminogen activator and PAI-1.* Ann Oncol, 2006. **17**(4): 597–604.
71. Heimann, R., et al., *Assessment of intratumoral vascularization (angiogenesis) in breast cancer prognosis.* Breast Cancer Res Treat, 1998. **52**(1–3): 147–58.
72. Blackwell, K.L., et al., *HER-2 gene amplification correlates with higher levels of angiogenesis and lower levels of hypoxia in primary breast tumors.* Clin Cancer Res, 2004. **10**(12 Pt 1): 4083–88.
73. Yen, L., et al., *Heregulin selectively upregulates vascular endothelial growth factor secretion in cancer cells and stimulates angiogenesis.* Oncogene, 2000. **19**(31): 3460–69.
74. Yang, W., et al., *ErbB2 overexpression correlates with increased expression of vascular endothelial growth factors A, C, and D in human breast carcinoma.* Cancer, 2002. **94**(11): 2855–61.
75. Finkenzeller, G., et al., *Activated Neu/ErbB-2 induces expression of the vascular endothelial growth factor gene by functional activation of the transcription factor Sp 1.* Angiogenesis, 2004. **7**(1): 59–68.
76. Klos, K.S., et al., *ErbB2 increases vascular endothelial growth factor protein synthesis via activation of mammalian target of rapamycin/p70S6K leading to increased angiogenesis and spontaneous metastasis of human breast cancer cells.* Cancer Res, 2006. **66**(4): 2028–37.
77. Yang, W., et al., *ErbB2 overexpression in human breast carcinoma is correlated with p21Cip1 up-regulation and tyrosine-15 hyperphosphorylation of p34Cdc2: poor responsiveness to chemotherapy with cyclophoshamide methotrexate, and 5-fluorouracil is associated with Erb2 overexpression and with p21Cip1 overexpression.* Cancer, 2003. **98**(6): 1123–30.
78. Traxler, P., et al., *AEE788: a dual family epidermal growth factor receptor/ErbB2 and vascular endothelial growth factor receptor tyrosine kinase inhibitor with antitumor and antiangiogenic activity.* Cancer Res, 2004. **64**(14): 4931–41.
79. Mackay, A., et al., *cDNA microarray analysis of genes associated with ERBB2 (HER2/neu) overexpression in human mammary luminal epithelial cells.* Oncogene, 2003. **22**(17): 2680–88.
80. Bon, G., et al., *Loss of beta4 integrin subunit reduces the tumorigenicity of MCF7 mammary cells and causes apoptosis upon hormone deprivation.* Clin Cancer Res, 2006. **12**(11 Pt 1): 3280–87.
81. Yoon, S.O., S. Shin, and E.A. Lipscomb, *A novel mechanism for integrin-mediated ras activation in breast carcinoma cells: the alpha6beta4 integrin regulates ErbB2 translation and*

transactivates epidermal growth factor receptor/ErbB2 signaling. Cancer Res, 2006. **66**(5): 2732–39.

82. Hannigan, G.E., et al., *Regulation of cell adhesion and anchorage-dependent growth by a new beta 1-integrin-linked protein kinase.* Nature, 1996. **379**(6560): 91–96.
83. Wu, C., et al., *Integrin-linked protein kinase regulates fibronectin matrix assembly, E-cadherin expression, and tumorigenicity.* J Biol Chem, 1998. **273**(1): 528–36.
84. Xie, W., et al., *Expression of the integrin-linked kinase (ILK) in mouse skin: loss of expression in suprabasal layers of the epidermis and up-regulation by erbB-2.* Am J Pathol, 1998. **153**(2): 367–72.
85. Khoury, H., et al., *HGF converts ErbB2/Neu epithelial morphogenesis to cell invasion.* Mol Biol Cell, 2005. **16**(2): 550–61.
86. Price-Schiavi, S.A., et al., *Expression, location, and interactions of ErbB2 and its intramembrane ligand Muc4 (sialomucin complex) in rat mammary gland during pregnancy.* J Cell Physiol, 2005. **203**(1): 44–53.
87. Pino, V., et al., *Membrane mucin Muc4 induces density-dependent changes in ERK activation in mammary epithelial and tumor cells: role in reversal of contact inhibition.* J Biol Chem, 2006. **281**(39): 29411–20.
88. Chiodoni, C., et al., *Triggering CD40 on endothelial cells contributes to tumor growth.* J Exp Med, 2006. **203**(11): 2441–50.
89. Muller, A., et al., *Involvement of chemokine receptors in breast cancer metastasis.* Nature, 2001. **410**(6824): 50–56.
90. Nagasawa, T., et al., *Molecular cloning and characterization of a murine pre-B-cell growth-stimulating factor/stromal cell-derived factor 1 receptor, a murine homolog of the human immunodeficiency virus 1 entry coreceptor fusin.* Proc Natl Acad Sci USA, 1996. **93**(25): 14726–29.
91. Braun, S., et al., *ErbB2 overexpression on occult metastatic cells in bone marrow predicts poor clinical outcome of stage I-III breast cancer patients.* Cancer Res, 2001. **61**(5): 1890–95.
92. Li, Y.M., et al., *Upregulation of CXCR4 is essential for HER2-mediated tumor metastasis.* Cancer Cell, 2004. **6**(5): 459–69.
93. Cabioglu, N., et al., *Chemokine receptor CXCR4 expression in breast cancer as a potential predictive marker of isolated tumor cells in bone marrow.* Clin Exp Metastasis, 2005. **22**(1): 39–46.
94. Corsini, C., et al., *Stroma cells: a novel target of herceptin activity.* Clin Cancer Res, 2003. **9**(5): 1820–25.

Chapter 5
Convergance of Cytoskeletal Signaling at p21-Activated Kinases

Anupama E. Gururaj and Rakesh Kumar

Abstract The PAK family of kinases regulates many aspects of cellular responses to external stimuli including cell migration. This process depends on organization of the actin cytoskeleton into adhesive and protrusive organelles in response to extracellular signals. PAKs are important nodes for the spatiotemporal control of actin-based motility in higher eukaryotes. PAKs are also central elements of signaling pathways that provide influence over virtually every major cellular function, including cell survival, cell differentiation and cell proliferation. This review depicts the roles of PAKs in cell migration and discusses how PAKs integrate with other sub-cellular systems involved in cell motility. Importantly, we also present an overview of the diverse functions of PAKs in the normal and pathological contexts. Our review concludes with a discussion of the lacunae and the future directions in the field of PAK biology.

Keywords P21 activated kinase · Adhesion · Cytoskeleton · Gene regulation · Integrins · Hergulin · PI3 kinase · Epithelial-to-Mesenchimal Transition (EMT) · Myosin light chain · Dynein light chain

5.1 Introduction

Controlled and organized response to external cues is a survival requirement for cells. Cell interactions with the extracellular matrix (cell–matrix adhesion) and neighboring cells (cell–cell adhesion) contribute to cell fate (proliferation, survival and differentiation) and behavior (polarization, phagocytosis, motility, metastasis, chemotaxis, and cytokinesis among others, see [1]). Cell migration is a fundamental process in the multicellular organism and regulated migration of cells underlies epithelial turnover and regeneration processes, such as in skin and wound healing [2, 3]. Cell motility can also take place under pathological conditions, such as in

R. Kumar
Department of Molecular and Cellular Oncology, The University of Texas, MD Anderson Cancer Center, Houston, TX 77030, USA
e-mail: rkumar@mdanderson.org

M. Bar-Eli (ed.), *Regulation of Gene Expression in the Tumor Environment*,

inflammatory diseases and cancer [4]. During oncogenic transformation, persisting cell proliferation is frequently coupled to the early onset of cancer cell motility, resulting either from the loss of cell–cell junctions, altered intracellular signaling or as a consequence of pro-migratory factors, such as chemokine gradients or growth factors produced by the tumor microenvironment [5]. Cell migration requires continuous assembly and disassembly of cell-ECM or cell–cell contacts, and constant remodeling of the associated actin cytoskeleton (Fig. 5.1). Communication and interaction between adhesion receptors and polymerized actin is therefore essential for important cell responses, such as migration and intracellular signaling that controls cell growth and survival. Thus, understanding how cells move entails insights into the dynamic of disassembly, relocation and reassembly of adhesion-associated cytoskeletal structures within the cell. Considerable effort has been directed to linking the cascade of signals from transmembrane receptors to downstream effectors of the actin cytoskeleton. These efforts have shown that the Rho GTPases, Cdc42, Rac, and Rho, are principal targets that promote distinct cytoskeletal changes leading to the formation of filopodia, lamellipodia or stress fibers, respectively [6–8]. Tightly controlled functional interactions between this family of small GTPases and their effectors allow for regulation of actin dynamics that culminates in cell motility.

Basic mechanisms of cell migration were first established in fibroblasts or keratinocytes, and later confirmed for cancer cells. The migratory machinery executes five interdependent steps, which are repeated in a cyclical manner [4] (Fig. 5.1). First, pseudopodia protrusion and polarization are initiated by actin polymerization in an adhesion-independent manner. Next, cell protrusions adhere to extracellular matrix (ECM) via adhesion molecules, thereby forming one or several focal adhesion sites. If expressed at the cell surface, proteases may become recruited to attachment sites and mediate the proteolytic cleavage of matrix barriers. Driven by actomyosin contraction of the cell body, bipolar tension develops towards the tissue and along the cell body, ultimately leading to the shortening of the cell length axis

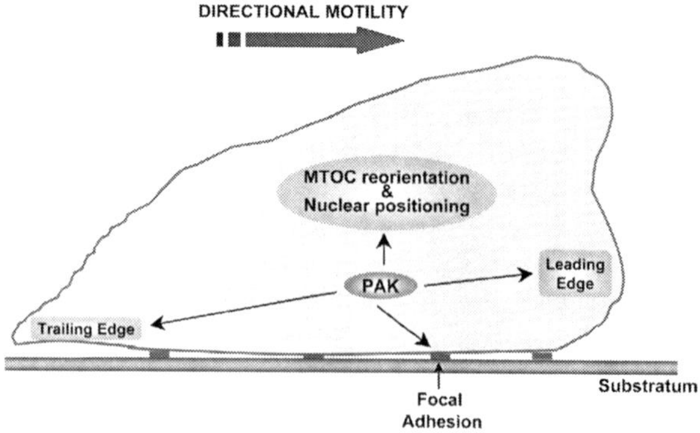

Fig. 5.1 A schematic representation of directional cell motility on the ECM. Cycles of cell–ECM attachment and detachment participate in cell motility and these are coordinated with the spatial regulation of the actin cytoskeleton. The figure shows the specific processes in which PAK1 is involved

and gradual gliding of the rear end. The above five-step model of cell migration was confirmed for tumor cells and different two-dimensional (2D) and three-dimensional (3D) ECM substrates [9–12].

For cell adhesion to ECM structures, adhesion molecules, such as those of the integrin family, cluster at the emerging adhesion site and recruit intracellular signaling and adaptor molecules. Adaptor molecules, such as alpha-actinin, talin, focal adhesion kinase (FAK), tensin, paxillin and vinculin connect integrins to the filamentous actin cytoskeleton forming a focal contact or focal adhesion (Fig. 5.1) [13]. After ECM binding, integrins cause phosphorylation and dephosphorylation events of regulatory molecules and downstream signaling cascades into the cell, termed 'outside-in signaling' [14]. Depending on the nature of ECM and intracellular signals, focal contacts can mediate dynamic cell behavior resulting in forward gliding of the cell body, or prompt cell adhesion and stable arrest.

Focal contacts further facilitate the recruitment of matrix proteases towards ECM substrate and the onset of peri-cellular proteolysis. Proteases are either membrane-tethered, such as the membrane-type matrix metalloproteases (MT-MMP), or serine proteases, such as urokinase-type plasminogen activator (uPA) or seprase, or are secreted and become bound to membrane receptors. MMP-1 specifically interacts with a2b1 integrin and thereby becomes recruited to points of cell contact with collagen I [15]. MT1-MMP colocalizes with clustered β1 or β3 integrins, thereby gaining access to cleave ECM substrate, such as collagen fibers [16,17]. Consequently, cells invading 3D tissue generate proteolytic degradation tracks bordered by proteolytically processed ECM [18]. In order for the cell to create tension and to change shape, actin contractility is mediated by myosin II, which is regulated by phosphorylation events of its myosin light chain (MLC). Phosphorylation is controlled by myosin light chain kinase (MLCK) and upstream effectors of MLCK such as PAK1 (Fig. 5.2). Therefore, the dynamic regulation of these molecules contributes to the maintenance of the migratory phenotype [19]. In tumor cells, this basic cell migration machinery is retained and most of the times, become 'overfunctional' by either upregulation or ongoing activation of integrins, proteases and cytoskeletal regulators.

In addition to signals transduced to actin and beyond, ligand-bound adhesion molecules can induce traction on attached filamentous actin, which results in adhesion receptor clustering and concentration of particular signaling complexes into localized domains at the membrane. Thus, bidirectional signaling occurs between the actin cytoskeleton and adhesion molecules at the plasma membrane and this coordinates and controls actin filament assembly, cell adhesion and adhesion-dependent signaling. Like assembly, localized disassembly of adhesive interactions is a critical part of adhesion dynamics and cell migration. This too is complex and involves multiple signaling inputs. Importantly, both receptor and non-receptor kinases contribute to adhesion turnover and the mechanisms are being identified by which kinases, and their effectors, suppress adhesive function and induce disorganization, or reorganization, of the associated actin cytoskeleton. Often this involves antagonism of the mediators of actin assembly. In addition, more direct mechanisms of inducing de-adhesion also occur. Indeed, the exact role of the kinases

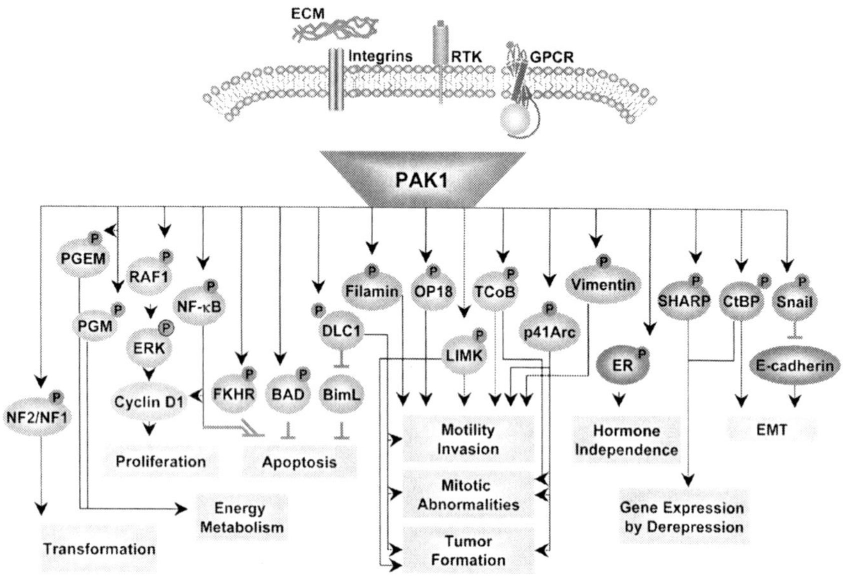

Fig. 5.2 Depiction of activation routes for PAK1 and downstream substrates of PAK1 that have a role in tumorigenesis. Lines with arrows depict stimulatory pathways while lines with crossbars indicate inhibitory action. The physiological outcomes in which the phosphorylated substrates play a role are shown in *yellow* boxes

is complex and is spatially regulated. Thus, such kinases simultaneously control actin/adhesion dynamics and signal into the cell interior to link proliferation with adhesion dynamics. As kinases are found at both adhesion types, and are profoundly important as modulators of both the adhesions themselves and the associated actin cytoskeleton, we will emphasize the close, but complex, three-way relations between the mediators of cell–ECM adhesions (the integrins), or cell–cell adhesions (the cadherins), kinases and regulators of the actin cytoskeleton with specific focus on one of the downstream effectors of Rho GTPases, the serine-threonine kinase p21-activated kinase (PAK) family.

5.2 The PAK Family

Originally discovered as a serine/threonine protein kinase activated by Cdc42 and Rac1 (hence the name, p21-activated kinase), the PAK family has assumed proportions beyond its initially perceived role in cytoskeletal remodeling events. More than a decade earlier, PAK was identified as a Cdc42 and Rac1 interacting serine/threonine kinase protein from rat brain that was activated after binding to the p21 proteins [20]. Subsequent years saw an explosion of information on PAKs and the PAK family grew to accommodate five more members. Advances have also been made with reference to the structure and biochemistry of the PAK family.

(Please see [21] for a more extensive explanation.) The PAK family is classified into two major groups – Group I and II based on the structural motifs found in the members. Group I PAKs include PAKs 1-3 while Group II encompasses PAKs 4-6. In general, all of these kinases consist of an amino-terminal p21 binding domain (PBD) that is followed by the kinase domain at the carboxyl terminus. Other functional motifs such as SH2 and SH3 interaction motifs are also found distributed across the molecule. The PBD also harbors an autoinhibitory region that reduces the kinase activity of PAKs. There is high degree of homology in the amino acid sequence specifically in the kinase domain between the members of the family; the conservation is higher between members of the subgroups. By virtue of earlier identification, Group I PAKs have been more investigated as compared to Group II PAKs but emerging data suggests that different members are tightly regulated and serve unique, non-overlapping functions in the cell. Although interesting, a more extensive discussion of this topic is outside the scope of this review and can be found elsewhere [22].

5.3 The PAK Network

Like most kinases, the activity of PAKs is tightly regulated in the cell. The kinase activity of PAKs is enhanced after Cdc42 or Rac1 binding [20]. The GTPases bind to PAKs only in an activated, GTP-bound form and the interaction occurs via the PBD of the PAKs [23]. Elucidation of the solution structure of PBD of PAK1 bound to Cdc42 and the crystal structure of PAK1 in its resting and activated states shed light on the exact mechanism by which the PAKs could be activated in the cells in response to extracellular stimuli [24, 25]. PAK1 exists as a dimer in its inactive conformation wherein the autoinhibitory domain acts like a flap to cover the kinase domain. The N-terminal region is involved in dimerization with another PAK molecule. Binding with activators induces a conformational change that results in a shift opening the catalytic cleft exposing key residues that then get autophosphorylated and activates the kinase [25]. Upstream signaling molecules such as Akt, PI3 kinase and PKA can also activate PAKs by trans-phosphorylating on key residues [26–28]. Likewise, when adaptor molecules like Grb2 and Nck get activated by growth factor signaling [29,30], they bind to PAK and release the conformational restraint to activate it. Further, nucleotide exchange factors like PIX are also capable of activating PAKs [31]. In addition, some lipids, especially sphingosine, can activate PAK1 independently of the GTPases [32]. Protein oligomerization also regulates the activity of PAKs, usually through release of the inhibitory constraint by binding to the autoinhibitory domain.

Activation of PAKs in the cells by various upstream regulators has been extensively investigated in the past but the role of negative regulators has been so far not well documented. Recent evidence indicates that PAKs are also subject to negative regulation. A few protein inhibitors of PAKs specifically PAK1 have been identified recently. These inhibit PAK1 by either of these two mechanisms – they bind to PAK1 and keep the inhibitory conformation intact (as seen in the cases of Merlin, Maspin, CRIPak) [33–35] or they act on the phosphorylated residues and dephosphorylate them to inactivate the PAK1 molecule (as seen in POX2/POX3 and hPIP1) [36, 37]. The importance of regulation by the endogenous inhibitors cannot be overstated since they keep the activity of PAKs confined to the appropriate timeframe and location. These could also be potential therapeutic targets in diseases that have deregulated PAK activity.

5.4 Physiology of PAK1

Analysis of normal functions of PAK1 reveals the critical role it plays in cytoskeletal reorganization and directional cell motility. Based on the homology of PAKs to the yeast Ste20, a known regulator of MAPK pathway and since the PAKs operated downstream of Cdc42 and Rac1, the first function of PAK to be explored was its role in directional motility (Fig. 5.1) and activation of JNK pathway. An elegant study demonstrated the formation of filopodia and membrane ruffles upon microinjection of activated PAK1 into fibroblasts [38]. The same group then went on to show conclusively that PAK1 had a definitive function in directional motility of fibroblasts [39]. Subsequent work confirmed regulation of motility by PAKs in other cell types including endothelial and cancer cells [40]. Numerous studies have fortified PAKs as important regulatory kinases in actin reorganization in various cellular backgrounds. Endothelial cell permeability is controlled by PAK via an effect on cell contractility [41]. Likewise, PAKs influence both smooth as well as cardiac muscle contraction through a variety of substrates. PAK1 phosphorylates MLCK and MLCP in smooth muscle cells [42] whereas it targets Troponin I, PP2A and desmin in cardiac muscles to bring about contraction [43]. Analysis of cytoskeleton remodeling in vascular smooth muscle also revealed that PAK along with its interaction partner PIX regulates podosome formation and consequently motility [44]. Surprisingly, ephrinA1 stimulated the PAK pathway to inhibit integrin-induced vascular smooth muscle cell spreading, indicating a complex role for PAK in this model system [45]. In human platelets, PAK supports a motile phenotype by triggering shape change through binding to cortactin [46], further reinforcing the concept that PAK plays a very important role in cytoskeletal remodeling.

Since PAKs control cell motility, it is highly probable that they have a role in embryogenesis where the proper directional motion of cells is of paramount importance. Indeed, PAKs have been demonstrated to have important functions in developmental processes across different organisms. In the process of axon extension, PAK1 is found associated with a complex (comprising netrin-1 receptor, Cdc42, Rac1 and N-wasp) that supports growth cone extension via formation of lamellipodia [47]. PAK1 is also involved in regulation of filopodia formation during axon development [48]. The Xenopus homolog of PAK5 is actively engaged in movement of cells during gastrulation by preventing calcium induced cell adhesiveness [49]. In a recent investigation, it was demonstrated that PAK1 downregulates formation of cell–cell adhesion in response to EphA4 signaling in Xenopus embryo for which the GTPase binding domain of PAK1 was sufficient but kinase activity was dispensable [50]. The role of PAKs in different develeopmental events in Drosophila is well studied with PAKs being implicated in photoreceptor cell morphogenesis [51] and synapse development [52]. Similarly, the C. elegans homolog of PAK affects embryo elongation and hypodermal cell fusion [53].

Another notable cellular process that is regulated by PAK1 directly is the production of ROS via NADPH Oxidase in neutrophils. The knowledge that Rac1 activates superoxide production by NADPH oxidase in neutrophils initiated a search for the signaling pathway involved in the process and led to the identification of PAK1 as a Rac1 binding molecule in these cells [54]. Subsequent work defined the specific role of PAK1 in neutrophils by elucidating that PAK1 could directly phosphorylate p67(phox), a component of NADPH oxidase [55]. This function assumes importance in the context of the tumor microenvironment where the control of the redox status determines the survival advantage conferred to the cell. Further, PAK1 associates with scaffolding proteins such as Grb2 to link EGFR signaling and downstream pathways [29]. PAK1 also associates with the scaffolding molecule, Nck [30, 56] to link upstream signaling events to activation of pathways necessary for eliciting the physiological response. Recent evidence implies a role for Rac1 and PAK1 in coordinating growth factor and integrin signalling. Specifically, adhesion induces re-localisation of Rac1 to the membrane, activation of PAK1 [57] and subsequent signaling to ERK by stabilizing MEK1/ERK complexes [58]. Recent evidence indicates that adhesion-dependent phosphorylation of PAK1, and then MEK1, on particular serine residues requires the activities of Src and FAK [59], indicating that adhesion- and mitogen-induced signaling converge at the level of PAK1. Thus, extensive investigations by several groups have elucidated the functional role played by PAKs in the physiological context.

Since majority of the focus has been on cytoskeletal functions of PAK1, other, more important functions of PAK1 could potentially have been overlooked. Recent studies bear witness to this observation. The phenomena of auto-inhibition and relocalization of signaling proteins to specific subcellular compartments play an important role in the initiation and regulation of signaling, which only recently has become more appreciated. An investigation revealed that PAK1 could translocate into the nucleus in response to growth factor signaling. The study identified nuclear localization signals in PAK1 and looked at the chromatin targets of PAK1 [60]. Thus, PAK1 can itself function as a transcriptional regulator to control gene expression in a specific situation.

The functions of PAK1 that have been elucidated thus far have been via identification of its binding partners and substrates (Fig. 5.2). Thus, role of PAK in motility was recognized due to its ability to bind small GTPases and subsequently, phosphorylate molecules like filamin, myosin light chain etc. In order to discover novel functions of PAKs, it is important to identify and study new binding partners and substrates of PAKs. Therefore, a deeper understanding of the downstream effectors is a novel approach to reveal the full scope of PAKs in human cancers specifically with reference to downstream effectors that have a role in cancer (Fig. 5.2).

5.5 PAKS in Cancer

Cancer cells exhibit altered growth properties and survival, as well as aberrant cytoskeletal organization associated with epithelial/mesenchymal plasticity and cell migration [61]. The involvement of integrin and cadherin changes, receptor tyrosine kinases like EGF receptor and ErbB-2, the Src and Abl families of non-receptor tyrosine kinases, as well as downstream effectors including FAK in cancer is well documented [62–69]. In addition, many of the Rho GTPases, their regulators and downstream effectors are also implicated in cancer development [70,71]. Likewise, there are many examples of altered expression and/or activity of PAK1, a component of the complex circuitry of signaling in cancer cells, and dysfunction of PAK1 is associated with the malignant phenotype [40]. So although the mechanisms involved in misregulation of the actin nucleation apparatus are not yet fully understood, its fundamental nature and dysfunction of upstream regulators implies that there may also be disruption of these pathways during tumourigenesis. Since PAKs are implicated in control of moltility, cell proliferation and survival, the role of this family in cancer is now beginning to be investigated.

An early clue for a role for PAKs in cancer came with the observation that the PBD region of PAK1 inhibited Ras and Rac-induced transformation, implying that this pathway is involved in tumorigenesis [72]. Cocomitantly, studies with the kinase-dead mutant of PAK1 revealed that the kinase activity of PAK1 was important for Ras-induced but not Raf-induced transformation of fibroblast cells [73]. Later investigations demonstrated that PAK1 was essential for transformation induced not only by Ras and Rac1 but also Rac3, vav3, and cdc42 [74,75]. Studies revealed a definitive role for PAK1 in motile and invasive phenotype observed in breast cancer cells. Heregulin-mediated motility and invasiveness seen in breast cancer cells required PAK1 activity [76]. Activation of PAK1 by heregulin stimulated its relocalization to the leading edges of cells. Heregulin-induced cytoskeletal reorganization and cell invasion of breast cancer cells was effectively blocked by an inhibitor of PI3-kinase, dominant-negative PAK1 mutants, or anti-HER2 antibody Herceptin [76]. Interestingly, both phosphatidylinositol 3-kinase and PAK1 are co-hyperactivated in breast tumors. In addition, during the course of an investigation looking into the expression and activity of PI3 kinase in breast cancers, it was also unearthed that PAK1 was overexpressed and hyperactivated giving the indication that PAK1 expression and activity could be deregulated in human tumors [77]. Subsequent studies focused on elucidating various functions of PAK1 by looking at its substrates in breast cancer cells. An elegant genetic study then showed that overexpression of constitutively active PAK1 in mouse mammary gland resulted in formation of hyperplastic nodules [78], minimal intraductal neoplasia and finally adenocarcinoma [79], providing direct evidence for PAK1 for a role in breast tumorigenesis.

Recent findings in a number of cancers indicate that PAK1 expression is increased in tumors, and that, in general, higher-grade tumors express higher levels of PAK1 protein. Analysis of PAK1 expression in human colorectal carcinoma biopsy samples demonstrated that PAK1 expression levels increased with tumour grade [80], with correlation between PAK1 expression and disease progression. PAK1 as well as PAK2 have been implicated in neurofibromatosis type 2 (NF2). Early investigations revealed that PAK2 could phosphorylate the NF2 tumor-suppressor gene product Merlin on Ser518, a site that affects its activity and localization [81]. Recently, it was also reported that merlin itself could inhibit the activation of PAK1 by binding to the PBD and inhibiting PAK1 recruitment to focal adhesions [33]. PAK1 deregulation is also observed in other tumors of neuronal origin. PAK1 expression levels are also increased in human breast tumors, with increased expression in invasive regions of the tumours correlating with increased cyclin D1 expression [82]. Recent findings also strongly indicate that PAK1 activation is necessary for lysophosphatidic acid-induced and autotoxin-induced cell motility in melanoma cells [83]. In addition to studies showing PAK1 function is required for transformation, PAK1 has also been demonstrated to be amplified in tumors. In ovarian cancer, PAK1 locus was amplified and correlated with cyclin D1 expression [84]. In breast tumors, PAK1 expression also showed positive correlation with cyclin D1, with nuclear localization of PAK1 playing a role in tamoxifen resistance [85]. While PAK1 is the most well studied member of the PAK family so far, attention is now being paid to other members of the family. Analysis of the expression of PAK family members indicates that overexpression as well as amplification of the other members as well was associated with various tumor types. Amplification of PAK4 locus has been identified by CGH analysis in pancreatic cancer [86]. In contrast to the other members, PAK2 seems to act as an inhibitor of tumor growth. PAK2 was recently demonstrated to be a negative regulator of myc protein, suggesting that PAK2 could potentially be used for therapy to inhibit myc-induced neoplasia [87].

These findings that PAK protein levels are increased in several types of cancer indicate that PAK gene transcription, translation or protein stability could be dysregulated in cancer cells. At present, little is known about the mechanisms of regulation of PAK transcription; however, regulators of PAK activity have been well defined.

Evidence for increased expression of PAKs in tumors is still limited; however, numerous studies have shown that PAK activity is a necessity for tumor formation and progression. This logically leads us to look closer into the events occurring downstream of PAK, namely its substrates and their role in cancer.

5.6 Downstream Effectors/Substrates of PAKs in Tumors

In general, tumors have been shown to have increased kinase activity as compared to normal tissue [40]. This implies that the substrates of PAKs are the driving force to bring about the biological effects of PAK1 signaling (Fig. 5.2). Thus, a discussion about substrates of PAK specifically those that could potentially have a role in cancer are relevant.

5.6.1 Influence of PAK1 on Actin Dynamics and EMT

As mentioned earlier, the first function of PAK1 to be elucidated was its role in cytoskeletal reorganization and thus, by implication, motility. It was not very long before the substrate specificity of PAK1 was identified based on sequence similarity to myosin heavy chain kinase [88]. The same investigation also identified myosin as a substrate, which was activated upon phosphorylation by PAK1. Once the substrate specificity was known, a variety of substrates of PAK1 were identified. An elegant study showed that PAK1 affects the phosphorylation state of myosin light chain (MLC), thus linking this kinase to a molecule that directly affects cell movement [39]. Decreased phosphorylation of MLC was mediated through MLC kinase (MLCK) wherein PAK1 phosphorylated MLCK to inhibit its activity, consequently reducing the phosphorylation on MLC [89]. The other motor system in the cell is the dynein motor and PAK1 regulates certain functions of two components of the motor complex, dynein light chain 1 & 2 (DLC1) [90, 91]. PAK1 phosphorylation of DLC1 plays an important role in cell survival while DLC2 has been reported to be an interacting partner of PAK1. Besides regulation of the dynein and myosin motors, PAK1 is also involved in actin reorganization events. PAK1 exerts its effects through its substrates, LIM kinase (LIMK), p41-Arc, and filamin. PAK1 phosphorylates LIMK on a serine residue in the activation loop, thus increasing the activity of LIMK leading to higher phosphorylation of cofilin, thereby inactivating its F-actin-depolymerizing activity and leading to accumulation of actin filaments and aggregates [92]. Further, FLNa phosphorylation by PAK1 seems to be essential for PAK1-induced cytoskeletal reorganization. Interestingly, FLNa also stimulates PAK1 activity implying that the interaction is important for local activation of PAK1 [93]. A recent study reported p41-Arc, a subunit of the Arp 2/3 complex, which is involved in the actin nucleation process, as a direct substrate of PAK1 [94]. The phosphorylation was found to be important for directional motility of breast cancer cells. PAK1 also interact with scaffolding proteins such as PIX and paxillin that are instrumental in localization and activation of PAK1 to focal adhesions [31, 95]. New evidence indicates that PAK1 is constitutively activated in breast cancer cells, mislocalized to focal adhesion and the PAK localization to these structures via PIX is required for the maintenance of paxillin- and PIX-containing focal adhesions [96]. PAK1 also controls intermediate filament component, vimentin through phosphorylation, which results in decreased filament formation [97]. Microtubule reorganization plays a critical role not only in cell division but also in movement of the cells. Stathmin, also called oncoprotein 18 (OP18), is a microtubule destabilizing molecule which upon phosphorylation by PAK1 gets inactivated, thus stabilizing microtubules at the leading edge [98]. An interesting substrate of PAK1 is GEF-H1 that is a microtubule-localized Rho exchange factor whose activity is controlled by microtubule binding and release. PAK1 phosphorylation occurs in a domain important for inhibition of exchange activity and induces 14-3-3 binding to GEF-H1 leading to relocation of 14-3-3 to microtubules [99]. Tubulin cofactor B (TcoB), a molecule involved in formation of tubulin heterodimers, was shown to be phosphorylated by PAK1 and phosphorylation was necessary for normal functioning of TcoB [100]. The wide range of molecules that PAK1 regulates during cytoskeletal reorganization supports the notion that PAK1 acts as a 'master controller' in this process. A careful perusal of PAK substrates that are involved in motility indicates that many of them are deregulated in a variety of human cancers. Since PAK regulates the activity of these molecules, this suggests a mechanism through which PAK could produce a motile and invasive phenotype in cancer cells.

Induction of EMT frequently occurs during carcinoma invasion wherein polarized epithelial cells are converted into motile, invasive cells. This is accompanied by loss of cell–cell contact that is typically mediated by the cadherin family of proteins. Hepatocyte growth factor (HGF), the ligand for the Met receptor tyrosine kinase, is a potent modulator of epithelial-mesenchymal transition and activates PAK1 in epithelial cells to induce cell spreading [101]. Snail, a corepressor protein promotes EMT by repressing E-cadherin expression and has been identified as a key molecule in regulation of EMT [102, 103]. The corepressor activity of Snail depends on its sub-cellular localization and phosphorylation by PAK1 confines Snail to the nucleus and promotes its corepressor functions [104]. Another corepressor whose activity is controlled by PAK1 phosphorylation is CtBP that also represses E-cadherin gene expression [105], thus implicating PAK as a regulatory molecule that could potentially control EMT.

5.6.2 Redox Homeostasis and PAK1

Tumor cells have altered metabolic phenotype and redox state as compared to their normal counterparts, which gives them a survival advantage over normal cells. The redox balance in the cells is maintained by reactive oxygen species (ROS) and the reducing equivalents. ROS are derived in the cell primarily from the membrane-associated enzyme, NADPH oxidase while the reducing equivalents arise from metabolic pathways like pentose phosphate pathway (PPP). PAK1 plays a vital role in regulation of both metabolic pathways as well as NADPH oxidase. PAK1 phosphorylates p47 (phox) subunit of NADPH oxidase [106] while p67 (phox) is phosphorylated by PAK2 [55]. Phosphorylation stimulates membrane localization of the subunits and leads to activation of NADPH oxidase. In neutrophils, PAK1 phosphorylates and inhibits PGAM, an enzyme of the glycolytic pathway, resulting in a transient switch from glycolysis to metabolism through the pentose phosphate pathway and an increased cellular supply of NADPH [107]. PAK1 also phosphorylates and enhances the enzymatic activity of PGM, an important regulatory enzyme that serves as a link between the glycolytic pathway and sucrose catabolism [108]. Thus, PAK1 could potentially cause changes required for a specific shift towards the utilization of the pentose phosphate pathway in tumor cells. Another of PAK1 substrates is the corepressor molecule, C-terminal binding protein (CtBP), which upon phosphorylation gets inactivated, and no longer functions as a corepressor [105]. CtBP is a redox sensitive corepressor molecule and NADH bound CtBP is more repressive than the NAD associated molecule. PAK1 preferntially phosphorylates the NADH bound form indicating that redox status of the cell is important for its function [105]. Thus, PAK1 is involved in maintainance of the redox balance and is also affected in terms of activity with changes in the redox balance in the cell.

5.6.3 PAK1 in Cell Survival Signaling

A characteristic of tumor cells is their ability to divide uncontrollably. By virtues of its homology to the yeast Ste20, PAK1 was implicated in the MAPK signaling cascade. Various groups demonstrated that PAK1 could induce the activation of JNK/SAPK [109, 110] and that PAK1 had a permissive role in the ERK pathway [111]. Activation of ERK was shown to occur via MEK1 [112, 113], implicating PAK1 in survival signaling. Furthermore, Raf-1, the initial kinase in the MAPK cascade, is phosphorylated and activated by PAK1 [113, 114]. The elucidation of the regulation of MAPK signalling immediately precipitated an intensive search for downstream targets involved in cell survival. An early downstream target identified was the pro-apoptotic factor BAD, which was shown to be regulated by phosphorylation which inhibited its pro-apoptotic function [114]. Subsequently, BAD phosphorylation in response to HIV-1 Nef was shown to be independent of Akt [115]. Recent evidence demonstrates the presence of a scaffolding protein, ArgBP2gamma that provides docking sites for both Akt as well as PAK1 and promotes cell survival [116]. Results from a recent investigation also indicate that PAK1 phosphorylation of Raf-1 induces translocation of Raf-1 to mitochondria where it phosphorylates BAD and forms a complex with Bcl-2 [117]. Also, the discovery that PAK1 could phosphorylate and inactivate one of the members of the forkhead family of transcription factors, FKHR further strengthened the case for involvement of PAK1 in survival mechanisms [118]. PAK1-mediated phosphorylation of FKHR regulates its subcellular distribution: phosphorylated FKHR is maintained in the cytosol, and is thus unable to mediate transcription. As mentioned earlier, evidence for involvement of PAK1-DLC1 in survival pathways came with studies showing PAK1 phosphorylation of DLC1 [90]. PAK1 recognizes DLC1-BimL dimers, phosphorylates both the molecules and incapacitates the interaction of BimL with Blc-2; an event that would inactivate Bcl-2. Thus, multiple mechanisms exist through which PAK promotes cell survival and proliferation.

5.6.4 Role of PAK1 in Anchorage Independence of Tumor Cells

Growth of cells without the necessity of a substratum to adhere to is a characteristic unique to tumor cells. This feature is called anchorage-independent growth and is defined as a property exhibited by transformed cells. First clues for the role of PAK in anchorage-independent growth came from the observation that dominant negative PAK1 could block Ras induced transformation of fibroblast cells [73]. Epithelial cells overexpressing kinase-active mutant of PAK1 exhibited increased anchorage-independent growth of cells in soft agar in a preferential mitogen-activated protein kinase-sensitive manner [119]. A follow-up study from the same group showed that Etk/Bmx tyrosine kinase activates PAK1 by phosphorylation and while overexpression of Etk stimulated anchorage-independent growth as well as proliferation and invasion of epithelial cells, ectopic expression of the kinase-dead mutant showed no change [120]. This implied that PAK1 phosphorylation could potentially be important for Etk/Bmx induced cancerous phenotype. DLC1 that is involved in PAK-regulated cell survival also increased anchorage independent growth of non-invasive breast cancer cells upon overexpression and could form tumors in nude mice [90]. A substrate of PAK1 with known function in actin reorganization also showed an unexpected role in anchorage-independent growth. Ectopic expression of p41-Arc in epithelial cells increased their ability to grow on soft agar as well as develop tumors in nude mice, which was not observed in the mutants that could not be phosphorylated by PAK1 [94]. Together, these results support the role of PAK1 in cancer and reinforce the notion that the oncogenic properties of PAK1 are a manifestation of the properties of its substrates.

5.6.5 PAK1 and Cell Cycle Control

Control of the cell cycle is a vital part of the cell's replication machinery and disruption of this process is commonly seen in tumors. The first report that PAK1 had a role in cell cycle came with studies showing inhibition of G2/M progression by Xenopus PAK in Xenopus [121]. This study also demonstrated for the first time that PKA and PAK could function similarly; a concept that is being explored in more detail only now. Confirming evidence was derived from experiments conducted with yeast PAK homologues, Ste20 and Cla4, both of which were found to be important for cytokinesis and actin polarization [122, 123]. Further, Shk1, a yeast PAK protein, was found to be important for microtubule dynamics and localized to interphase microtubules and mitotic microtubule spindles, as well as to cell ends and septum-forming regions of fission yeast cells [124]. In higher eukryotes, ectopic overexpression of kinase-active PAK1 resulted in appearance of multiple spindle orientations due to abnormal organization of the spindles [125]. Further experiments on the human PAK1 indicated that PAK1 showed different localizations during cell cycle [126]. In interphase cells, a subset of PAK1 molecules were localized inside the nucleus, an observation that led to identification of other nuclear functions later on. PAK1 was localized on the centrosomes and/or metaphase plate in metaphase-to-anaphase cells and on the midbody in telophase cells. The same study also identified histone H3 as a PAK1 substrate and PAK1 phosphorylated histone H3 on Ser 10, a site that has been shown to be crucial for chromosome condensation and cell-cycle progression as well as transcriptional activation [126, 127], further implicating a regulatory role for PAK1 in cell cycle control. Interestingly, PAK1 itself can be activated by phosphorylation during cell cycle by Cdc2 [128, 129] indicating that PAK1 activation could be a means for control of specific cell events. Merlin, a protein that has been identified as PAK1 substrate as well as inhibitor, has been linked to events related to cell cycle control and shows differential localization during cell cycle progression [130]. However, the authors found that phosphorylation of merlin by PAK1 did not affect its nuclear localization in the interphase nucleus. The role played by phosphorylation in other phases of the cell cycle is yet to be clarified.

Overexpression of another recently identified substrate of PAK1, TcoB leads to formation of multiple microtubule organizing centers that was observed in a PAK1 overexpression situation as well [100].

5.6.6 Regulation of Gene Expression by PAK1

Signaling cascades that originate at the intracellular membrane normally bring about changes in the gene expression in the nucleus. MEK1 has been shown to regulate a variety of transcription factors and thus gene expression [131]. PAK1 activates MEK1 pathway by phosphorylating MEK1 directly [112] and controls expression of specific genes. Further, activation of NFκB, a well characterized transcriptional regulator, by multiple pathways such as Ras, Rac , Raf-1 an well as lipopolysaccharides was demonstrated to unequivocally require PAK1 in multiple cell types [132], implying PAK1 dependent gene regulation could occur through NFκB. As mentioned earlier, PAK1 also phosphorylates CtBP and regulates its corepressor activities, thus regulating gene expression [105]. Of special interest is regulation of nuclear receptors like estrogen receptor (ER) and androgen receptor (AR) by PAK1 and PAK6 respectively. PAK6 was found to interact with both ER as well as AR and was in fact originally identified through a yeast-two-hybrid screen using AR as a bait [133]. Further characterization of AR and PAK6 interaction revealed that PAK6 inhibits AR signaling by blocking nuclear translocation of activated AR [134]. PAK1 on the other hand augments ER transactivation functions by phosphorylating it [78]. This phosphorylation site, which is at ser305, has been linked to tamoxifen-resistance seen in breast cancer cells [135]. Furthermore, PAK1 transgenic mice show increased expression of ER target genes [78]. PAK1 phosphorylation of ER was also responsible for upregulation of cyclin D1 [82], which has been shown to play a key role in tumorigenesis. Recent work has shown that DLC1, a PAK1 substrate, interacts with ER and facilitates ER's transactivation functions through a novel coactivator [136]; evidence also indicated that DLC1 plays a significant role in tumor cells hypersensitivity to estrogen.

The ability of PAK members to interact with steroid hormonal receptors suggests that PAKs may play an important role in the cross-talk between steroid hormone receptors and growth factor signaling pathways, which is implicated in endocrine therapy. Lastly, PAK1 has been recently identified to possess nuclear localization signals (NLS) that in response to external stimuli facilitate nuclear translocation [60]. Investigation of a functional consequence for translocation into the nucleus led to the discovery of chromatin targets of PAK1. PAK1 could act in the nucleus to both activate (as seen with PFKM) and repress gene expression (NFAT1). These studies have opened new avenues for exploration of hitherto unsuspected nuclear functions of the PAK family.

Role of PAK family in cytoskeletal modeling also implicates them in cancer invasion and metastasis (see section on motility). However, evidence for a direct role of this family of kinases in both invasion and metastasis is not comprehensive. Studies indicated that heregulin stimulated invasion of breast cancer cells was dependent on PAK1 activity [76]. Further, Rac induced invasion of epithelial cells could be inhibited by blocking PAK1 signaling pathways [137]. A recent report demonstrates the blockade of migration of cancer cells by emodin by interference with the formation of Cdc42/Rac1 and the p21-activated kinase complex [138]. Taken together, these results argue for a case for PAK1 as a key molecule that exerts control over a wide range of physiological events and thus, could potentially function as an oncogene.

5.7 Future Perspectives

It is clear that coordination of signals from adhesion receptors and RTKs, often by kinases linking to the actin regulatory machinery, is at the heart of cellular responses to extracellular cues. One of the emerging themes is the coordination of actin/adhesion dynamics and signalling output from adhesion sites. There is clearly a complex, and probably reciprocal, relationship between kinases and regulation of actin/adhesion assembly and turnover. At the same time, these kinases are equipped to control a diverse range of biological functions, including transducing signals from both growth factor- and adhesion-receptors. They therefore have a clear role in coordinating organisation of the actin cytoskeleton with growth responses. The tight and complex control of signaling upstream and downstream of the intracellular kinases is testimony to the importance, and pivotal nature, of their activities. Together with their adhesion receptor partners, the kinases and actin regulators allow three-way relaying of information that control the basic properties of normal cells.

As mentioned in this review, and elsewhere, a great deal is now known about the functional interplay between adhesion receptors, kinases and actin modulators. However, there are still huge gaps in our knowledge that require continued investigation; in particular, how the temporal and spatial integration of signals from both adhesion receptors and kinases that control cell fate are regulated. Future progress in understanding the complexities of adhesion receptor biology, and the associated circuitry that links adhesion signaling to other cell responses requires multi-faceted approaches. In this regard, though, new technologies have opened up powerful new avenues of research. For example, the use of fluorescent probes and new microscopic methods (such as FRET and FLIM), coupled with the development of protein activity biosensors, will allow amazing visual insights into spatial regulation of particular proteins and protein complexes in real time. In addition, the arrival of technologies such as gene expression profiling, bioinformatics, proteomics and RNA interference, coupled to use of good cell model systems and genetically tractable model organisms, will reveal yet more molecular detail of the way in which cells respond to their environment.

We have little doubt that what will be revealed is yet further complexity, and perhaps the major challenge that lies ahead is handling and integrating the vast amount of new information. Nevertheless, daunting though the task, we are moving towards a complete understanding of how cell–environment interactions control multicellular organisms.

Acknowledgments This work was supported by NIH grants CA90970, CA65746, and CA80066 (to R.K.).

References

1. Penninger, J.M. & Crabtree, G.R. The actin cytoskeleton and lymphocyte activation. *Cell* **96**, 9–12 (1999).
2. Woodley, D.T. et al. Re-epithelialization. Human keratinocyte locomotion. *Dermatol. Clin.* **11**, 641–646 (1993).
3. Witte, M.B. & Barbul, A. General principles of wound healing. *Surg. Clin. North Am.* **77**, 509–528 (1997).
4. Friedl, P. & Wolf, K. Tumour-cell invasion and migration: diversity and escape mechanisms. *Nat. Rev. Cancer* **3**, 362–374 (2003).
5. Orimo, A. et al. Stromal fibroblasts present in invasive human breast carcinomas promote tumor growth and angiogenesis through elevated SDF-1/CXCL12 secretion. *Cell* **121**, 335–348 (2005).
6. Jaffe, A.B. & Hall, A. RHO GTPases: biochemistry and biology. *Annu. Rev. Cell Dev. Biol.* (2005).
7. Higgs, H.N. & Pollard, T.D. Regulation of actin polymerization by Arp2/3 complex and WASp/Scar proteins. *J Biol. Chem.* **274**, 32531–32534 (1999).
8. Ridley, A.J. et al. Cell migration: integrating signals from front to back. *Science* **302**, 1704–1709 (2003).
9. Clark, E.A., Golub, T.R., Lander, E.S. & Hynes, R.O. Genomic analysis of metastasis reveals an essential role for RhoC. *Nature* **406**, 532–535 (2000).
10. Maaser, K. et al. Functional hierarchy of simultaneously expressed adhesion receptors: integrin alpha2beta1 but not CD44 mediates MV3 melanoma cell migration and matrix reorganization within three-dimensional hyaluronan-containing collagen matrices. *Mol. Biol. Cell* **10**, 3067–3079 (1999).
11. Rabinovitz, I. & Mercurio, A.M. The integrin alpha6beta4 functions in carcinoma cell migration on laminin-1 by mediating the formation and stabilization of actin-containing motility structures. *J Cell Biol.* **139**, 1873–1884 (1997).
12. Sabeh, F. et al. Tumor cell traffic through the extracellular matrix is controlled by the membrane-anchored collagenase MT1-MMP. *J Cell Biol.* **167**, 769–781 (2004).
13. Miyamoto, S. et al. Integrin function: molecular hierarchies of cytoskeletal and signaling molecules. *J Cell Biol.* **131**, 791–805 (1995).
14. Hynes, R.O. Integrins: bidirectional, allosteric signaling machines. *Cell* **110**, 673–687 (2002).
15. Dumin, J.A. et al. Pro-collagenase-1 (matrix metalloproteinase-1) binds the alpha(2)beta(1) integrin upon release from keratinocytes migrating on type I collagen. *J Biol. Chem.* **276**, 29368–29374 (2001).
16. Wolf, K. et al. Compensation mechanism in tumor cell migration: mesenchymal-amoeboid transition after blocking of pericellular proteolysis. *J Cell Biol.* **160**, 267–277 (2003).
17. Ohuchi, E. et al. Membrane type 1 matrix metalloproteinase digests interstitial collagens and other extracellular matrix macromolecules. *J Biol. Chem.* **272**, 2446–2451 (1997).
18. Wolf, K. & Friedl, P. Functional imaging of pericellular proteolysis in cancer cell invasion. *Biochimie* **87**, 315–320 (2005).
19. Sahai, E. & Marshall, C.J. Differing modes of tumour cell invasion have distinct requirements for Rho/ROCK signalling and extracellular proteolysis. *Nat. Cell Biol.* **5**, 711–719 (2003).
20. Manser, E., Leung, T., Salihuddin, H., Zhao, Z.S. & Lim, L. A brain serine/threonine protein kinase activated by Cdc42 and Rac1. *Nature* **367**, 40–46 (1994).
21. Bokoch, G.M. Biology of the p21-activated kinases. *Annu. Rev. Biochem.* **72**, 743–781 (2003).

22. Jaffer, Z.M. & Chernoff, J. p21-activated kinases: three more join the PAK. *Int. J Biochem. Cell Biol.* **34**, 713–717 (2002).
23. Burbelo, P.D., Drechsel, D. & Hall, A. A conserved binding motif defines numerous candidate target proteins for both Cdc42 and Rac GTPases. *J. Biol. Chem.* **270**, 29071–29074 (1995).
24. Morreale, A. et al. Structure of Cdc42 bound to the GTPase binding domain of PAK. *Nat. Struct. Biol.* **7**, 384–388 (2000).
25. Lei, M. et al. Structure of PAK1 in an autoinhibited conformation reveals a multistage activation switch. *Cell* **102**, 387–397 (2000).
26. Tsakiridis, T., Taha, C., Grinstein, S. & Klip, A. Insulin activates a p21-activated kinase in muscle cells via phosphatidylinositol 3-kinase. *J. Biol. Chem.* **271**, 19664–19667 (1996).
27. Jiang, K. et al. Pivotal role of phosphoinositide-3 kinase in regulation of cytotoxicity in natural killer cells. *Nat. Immunol.* **1**, 419–425 (2000).
28. Tang, Y., Zhou, H., Chen, A., Pittman, R.N. & Field, J. The Akt proto-oncogene links Ras to PAK and cell survival signals. *J. Biol. Chem.* **275**, 9106–9109 (2000).
29. Puto, L.A., Pestonjamasp, K., King, C.C. & Bokoch, G.M. p21-activated kinase 1 (PAK1) interacts with the Grb2 adapter protein to couple to growth factor signaling. *J. Biol. Chem.* **278**, 9388–9393 (2003).
30. Voisin, L., Larose, L. & Meloche, S. Angiotensin II stimulates serine phosphorylation of the adaptor protein Nck: physical association with the serine/threonine kinases PAK1 and casein kinase I. *Biochem. J.* **341**(Pt 1), 217–223 (1999).
31. Manser, E. et al. PAK kinases are directly coupled to the PIX family of nucleotide exchange factors. *Mol. Cell* **1**, 183–192 (1998).
32. Bokoch, G.M. et al. A GTPase-independent mechanism of p21-activated kinase activation: regulation by sphingosine and other biologically active lipids. *J. Biol. Chem.* **273**, 8137–8144 (1998).
33. Kissil, J.L. et al. Merlin, the product of the Nf2 tumor suppressor gene, is an inhibitor of the p21-activated kinase, PAK1. *Mol. Cell* **12**, 841–849 (2003).
34. Odero-Marah, V.A. et al. Maspin regulates different signaling pathways for motility and adhesion in aggressive breast cancer cells. *Cancer Biol. Ther.* **2**, 398–403 (2003).
35. Talukder, A.H., Meng, Q. & Kumar, R. CRIPak, a novel endogenous PAK1 inhibitor. *Oncogene* **25**, 1311–1319 (2006).
36. Xia, C. et al. Regulation of the p21-activated kinase (PAK) by a human Gbeta -like WD-repeat protein, hPIP1. *Proc. Natl. Acad. Sci. USA* **98**, 6174–6179 (2001).
37. Koh, C.G., Tan, E.J., Manser, E. & Lim, L. The p21-activated kinase PAK is negatively regulated by POPX1 and POPX2, a pair of serine/threonine phosphatases of the PP2C family. *Curr. Biol.* **12**, 317–321 (2002).
38. Sells, M.A. et al. Human p21-activated kinase (PAK1) regulates actin organization in mammalian cells. *Curr. Biol.* **7**, 202–210 (1997).
39. Sells, M.A., Boyd, J.T. & Chernoff, J. p21-activated kinase 1 (PAK1) regulates cell motility in mammalian fibroblasts. *J. Cell Biol.* **145**, 837–849 (1999).
40. Kumar, R., Gururaj, A.E. & Barnes, C.J. p21-activated kinases in cancer. *Nat. Rev. Cancer* **6**, 459–471 (2006).
41. Stockton, R.A., Schaefer, E. & Schwartz, M.A. p21-activated kinase regulates endothelial permeability through modulation of contractility. *J. Biol. Chem.* **279**, 46621–46630 (2004).
42. Somlyo, A.P. & Somlyo, A.V. Ca2+ sensitivity of smooth muscle and nonmuscle myosin II: modulated by G proteins, kinases, and myosin phosphatase. *Physiol Rev.* **83**, 1325–1358 (2003).
43. Ke, Y., Wang, L., Pyle, W.G., de Tombe, P.P. & Solaro, R.J. Intracellular localization and functional effects of P21-activated kinase-1 (PAK1) in cardiac myocytes. *Circ. Res.* **94**, 194–200 (2004).
44. Webb, B.A. et al. PAK1 induces podosome formation in A7r5 vascular smooth muscle cells in a PIX-dependent manner. *Am. J. Physiol. Cell Physiol.* **289**, C898–907 (2005).

45. Deroanne, C., Vouret-Craviari, V., Wang, B. & Pouyssegur, J. EphrinA1 inactivates integrin-mediated vascular smooth muscle cell spreading via the Rac/PAK pathway. *J. Cell Sci.* **116**, 1367–1376 (2003).
46. Vidal, C., Geny, B., Melle, J., Jandrot-Perrus, M. & Fontenay-Roupie, M. Cdc42/Rac1-dependent activation of the p21-activated kinase (PAK) regulates human platelet lamellipodia spreading: implication of the cortical-actin binding protein cortactin. *Blood* **100**, 4462–4469 (2002).
47. Shekarabi, M. et al. Deleted in colorectal cancer binding netrin-1 mediates cell substrate adhesion and recruits Cdc42, Rac1, PAK1, and N-WASP into an intracellular signaling complex that promotes growth cone expansion. *J. Neurosci.* **25**, 3132–3141 (2005).
48. Kim, M.D., Kamiyama, D., Kolodziej, P., Hing, H. & Chiba, A. Isolation of Rho GTPase effector pathways during axon development. *Dev. Biol.* **262**, 282–293 (2003).
49. Faure, S. et al. Xenopus p21-activated kinase 5 regulates blastomeres' adhesive properties during convergent extension movements. *Dev. Biol.* **277**, 472–492 (2005).
50. Bisson, N., Poitras, L., Mikryukov, A., Tremblay, M. & Moss, T. Epha4 signaling regulates blastomere adhesion in the xenopus embryo by recruiting PAK1 to suppress Cdc42 function. *Mol. Biol. Cell* **18**, 1030–1043 (2007).
51. Schneeberger, D. & Raabe, T. Mbt, a Drosophila PAK protein, combines with Cdc42 to regulate photoreceptor cell morphogenesis. *Development* **130**, 427–437 (2003).
52. Albin, S.D. & Davis, G.W. Coordinating structural and functional synapse development: postsynaptic p21-activated kinase independently specifies glutamate receptor abundance and postsynaptic morphology. *J. Neurosci.* **24**, 6871–6879 (2004).
53. Chen, W., Chen, S., Yap, S.F. & Lim, L. The Caenorhabditis elegans p21-activated kinase (CePAK) colocalizes with CeRac1 and CDC42Ce at hypodermal cell boundaries during embryo elongation. *J. Biol. Chem.* **271**, 26362-26368 (1996).
54. Prigmore, E. et al. A 68-kDa kinase and NADPH oxidase component p67phox are targets for Cdc42Hs and Rac1 in neutrophils. *J Biol. Chem.* **270**, 10717–10722 (1995).
55. Ahmed, S. et al. Cryptic Rac-binding and p21(Cdc42Hs/Rac)-activated kinase phosphorylation sites of NADPH oxidase component p67(phox). *J. Biol. Chem.* **273**, 15693–15701 (1998).
56. Bokoch, G.M. et al. Interaction of the Nck adapter protein with p21-activated kinase (PAK1). *J. Biol. Chem.* **271**, 25746–25749 (1996).
57. del Pozo, M.A., Price, L.S., Alderson, N.B., Ren, X.D. & Schwartz, M.A. Adhesion to the extracellular matrix regulates the coupling of the small GTPase Rac to its effector PAK. *EMBO J.* **19**, 2008–2014 (2000).
58. Eblen, S.T., Slack, J.K., Weber, M.J. & Catling, A.D. Rac-PAK signaling stimulates extracellular signal-regulated kinase (ERK) activation by regulating formation of MEK1-ERK complexes. *Mol. Cell Biol.* **22**, 6023–6033 (2002).
59. Slack-Davis, J.K. et al. PAK1 phosphorylation of MEK1 regulates fibronectin-stimulated MAPK activation. *J. Cell Biol.* **162**, 281–291 (2003).
60. Singh, R.R., Song, C., Yang, Z. & Kumar, R. Nuclear localization and chromatin targets of p21-activated kinase 1. *J. Biol. Chem.* **280**, 18130–18137 (2005).
61. Hanahan, D. & Weinberg, R.A. The hallmarks of cancer. *Cell* **100**, 57–70 (2000).
62. Summy, J.M. & Gallick, G.E. Src family kinases in tumor progression and metastasis. *Cancer Metastasis Rev.* **22**, 337–358 (2003).
63. Felding-Habermann, B. Integrin adhesion receptors in tumor metastasis. *Clin. Exp. Metastasis* **20**, 203–213 (2003).
64. Gabarra-Niecko, V., Schaller, M.D. & Dunty, J.M. FAK regulates biological processes important for the pathogenesis of cancer. *Cancer Metastasis Rev.* **22**, 359–374 (2003).
65. McLean, G.W., Avizienyte, E. & Frame, M.C. Focal adhesion kinase as a potential target in oncology. *Expert. Opin. Pharmacother.* **4**, 227–234 (2003).
66. Menard, S., Pupa, S.M., Campiglio, M. & Tagliabue, E. Biologic and therapeutic role of HER2 in cancer. *Oncogene* **22**, 6570–6578 (2003).

67. Mendelsohn, J. & Baselga, J. Status of epidermal growth factor receptor antagonists in the biology and treatment of cancer. *J. Clin. Oncol.* **21**, 2787–2799 (2003).
68. Frame, M.C. Src in cancer: deregulation and consequences for cell behaviour. *Biochim. Biophys. Acta* **1602**, 114–130 (2002).
69. Pendergast, A.M. The Abl family kinases: mechanisms of regulation and signaling. *Adv. Cancer Res.* **85**, 51–100 (2002).
70. Malliri, A. & Collard, J.G. Role of Rho-family proteins in cell adhesion and cancer. *Curr. Opin. Cell Biol.* **15**, 583–589 (2003).
71. Sahai, E. & Marshall, C.J. RHO-GTPases and cancer. *Nat. Rev. Cancer* **2**, 133–142 (2002).
72. Osada, S., Izawa, M., Koyama, T., Hirai, S. & Ohno, S. A domain containing the Cdc42/Rac interactive binding (CRIB) region of p65PAK inhibits transcriptional activation and cell transformation mediated by the Ras-Rac pathway. *FEBS Lett.* **404**, 227–233 (1997).
73. Tang, Y. et al. Kinase-deficient PAK1 mutants inhibit Ras transformation of Rat-1 fibroblasts. *Mol. Cell Biol.* **17**, 4454–4464 (1997).
74. Mira, J.P., Benard, V., Groffen, J., Sanders, L.C. & Knaus, U.G. Endogenous, hyperactive Rac3 controls proliferation of breast cancer cells by a p21-activated kinase-dependent pathway. *Proc. Natl. Acad. Sci. USA* **97**, 185–189 (2000).
75. Sachdev, P., Zeng, L. & Wang, L.H. Distinct role of phosphatidylinositol 3-kinase and Rho family GTPases in Vav3-induced cell transformation, cell motility, and morphological changes. *J. Biol. Chem.* **277**, 17638–17648 (2002).
76. Adam, L. et al. Heregulin regulates cytoskeletal reorganization and cell migration through the p21-activated kinase-1 via phosphatidylinositol-3 kinase. *J. Biol. Chem.* **273**, 28238–28246 (1998).
77. Salh, B., Marotta, A., Wagey, R., Sayed, M. & Pelech, S. Dysregulation of phosphatidylinositol 3-kinase and downstream effectors in human breast cancer. *Int. J. Cancer* **98**, 148–154 (2002).
78. Wang, R.A., Mazumdar, A., Vadlamudi, R.K. & Kumar, R. P21-activated kinase-1 phosphorylates and transactivates estrogen receptor-alpha and promotes hyperplasia in mammary epithelium. *EMBO J.* **21**, 5437–5447 (2002).
79. Wang, R.A., Zhang, H., Balasenthil, S., Medina, D. & Kumar, R. PAK1 hyperactivation is sufficient for mammary gland tumor formation. *Oncogene* **25**, 2931–2936 (2006).
80. Carter, J.H. et al. PAK-1 expression increases with progression of colorectal carcinomas to metastasis. *Clin. Cancer Res.* **10**, 3448–3456 (2004).
81. Kissil, J.L., Johnson, K.C., Eckman, M.S. & Jacks, T. Merlin phosphorylation by p21-activated kinase 2 and effects of phosphorylation on merlin localization. *J. Biol. Chem.* **277**, 10394–10399 (2002).
82. Balasenthil, S. et al. p21-activated kinase-1 signaling mediates cyclin D1 expression in mammary epithelial and cancer cells. *J. Biol. Chem.* **279**, 1422–1428 (2004).
83. Jung, I.D. et al. Activation of p21-activated kinase 1 is required for lysophosphatidic acid-induced focal adhesion kinase phosphorylation and cell motility in human melanoma A2058 cells. *Eur. J. Biochem.* **271**, 1557–1565 (2004).
84. Schraml, P. et al. Combined array comparative genomic hybridization and tissue microarray analysis suggest PAK1 at 11q13.5-q14 as a critical oncogene target in ovarian carcinoma. *Am. J. Pathol.* **163**, 985–992 (2003).
85. Holm, C. et al. High PAK1 expression and nuclear localization is associated with lack of tamoxifen response in human premenopausal breast cancer and experimental models. *J. Natl. Cancer Inst.* **98**, 671–680 (2006).
86. Mahlamaki, E.H. et al. High-resolution genomic and expression profiling reveals 105 putative amplification target genes in pancreatic cancer. *Neoplasia* **6**, 432–439 (2004).
87. Huang, Z., Traugh, J.A. & Bishop, J.M. Negative control of the Myc protein by the stress-responsive kinase PAK2. *Mol. Cell Biol.* **24**, 1582–1594 (2004).
88. Brzeska, H., Knaus, U.G., Wang, Z.Y., Bokoch, G.M. & Korn, E.D. p21-activated kinase has substrate specificity similar to Acanthamoeba myosin I heavy chain kinase and activates Acanthamoeba myosin I. *Proc. Natl. Acad. Sci. USA* **94**, 1092–1095 (1997).

89. Sanders, L.C., Matsumura, F., Bokoch, G.M. & De Lanerolle, P. Inhibition of myosin light chain kinase by p21-activated kinase. *Science* **283**, 2083–2085 (1999).
90. Vadlamudi, R.K. et al. Dynein light chain 1, a p21-activated kinase 1-interacting substrate, promotes cancerous phenotypes. *Cancer Cell* **5**, 575–585 (2004).
91. Lu, J. et al. Identification of dynein light chain 2 as an interaction partner of p21-activated kinase 1. *Biochem. Biophys. Res. Commun.* **331**, 153–158 (2005).
92. Edwards, D.C., Sanders, L.C., Bokoch, G.M. & Gill, G.N. Activation of LIM-kinase by PAK1 couples Rac/Cdc42 GTPase signalling to actin cytoskeletal dynamics. *Nat. Cell Biol.* **1**, 253–259 (1999).
93. Vadlamudi, R.K. et al. Filamin is essential in actin cytoskeletal assembly mediated by p21-activated kinase 1. *Nat. Cell Biol.* **4**, 681–690 (2002).
94. Vadlamudi, R.K., Li, F., Barnes, C.J., Bagheri-Yarmand, R. & Kumar, R. p41-Arc subunit of human Arp2/3 complex is a p21-activated kinase-1-interacting substrate. *EMBO Rep.* **5**, 154–160 (2004).
95. Hashimoto, S., Tsubouchi, A., Mazaki, Y. & Sabe, H. Interaction of paxillin with p21-activated Kinase (PAK): association of paxillin alpha with the kinase-inactive and the Cdc42-activated forms of PAK3. *J. Biol. Chem.* **276**, 6037–6045 (2001).
96. Stofega, M.R., Sanders, L.C., Gardiner, E.M. & Bokoch, G.M. Constitutive p21-activated kinase (PAK) activation in breast cancer cells as a result of mislocalization of PAK to focal adhesions. *Mol. Biol. Cell* **15**, 2965–2977 (2004).
97. Goto, H. et al. Phosphorylation and reorganization of vimentin by p21-activated kinase (PAK). *Genes Cells* **7**, 91–97 (2002).
98. Daub, H., Gevaert, K., Vandekerckhove, J., Sobel, A. & Hall, A. Rac/Cdc42 and p65PAK regulate the microtubule-destabilizing protein stathmin through phosphorylation at serine 16. *J. Biol. Chem.* **276**, 1677–1680 (2001).
99. Zenke, F.T. et al. p21-activated kinase 1 phosphorylates and regulates 14-3-3 binding to GEF-H1, a microtubule-localized Rho exchange factor. *J. Biol. Chem.* **279**, 18392–18400 (2004).
100. Vadlamudi, R.K. et al. p21-activated kinase 1 regulates microtubule dynamics by phosphorylating tubulin cofactor B. *Mol. Cell Biol.* **25**, 3726–3736 (2005).
101. Royal, I., Lamarche-Vane, N., Lamorte, L., Kaibuchi, K. & Park, M. Activation of cdc42, rac, PAK, and rho-kinase in response to hepatocyte growth factor differentially regulates epithelial cell colony spreading and dissociation. *Mol. Biol. Cell* **11**, 1709–1725 (2000).
102. Barrallo-Gimeno, A. & Nieto, M.A. The Snail genes as inducers of cell movement and survival: implications in development and cancer. *Development* **132**, 3151–3161 (2005).
103. Come, C., Arnoux, V., Bibeau, F. & Savagner, P. Roles of the transcription factors snail and slug during mammary morphogenesis and breast carcinoma progression. *J. Mammary Gland Biol. Neoplasia* **9**, 183–193 (2004).
104. Yang, Z. et al. PAK1 phosphorylation of snail, a master regulator of epithelial-to-mesenchyme transition, modulates snail's subcellular localization and functions. *Cancer Res.* **65**, 3179–3184 (2005).
105. Barnes, C.J. et al. Functional inactivation of a transcriptional corepressor by a signaling kinase. *Nat. Struct. Biol.* **10**, 622–628 (2003).
106. Knaus, U.G., Morris, S., Dong, H.J., Chernoff, J. & Bokoch, G.M. Regulation of human leukocyte p21-activated kinases through G protein – coupled receptors. *Science* **269**, 221–223 (1995).
107. Shalom-Barak, T. & Knaus, U.G. A p21-activated kinase-controlled metabolic switch up-regulates phagocyte NADPH oxidase. *J. Biol. Chem.* **277**, 40659–40665 (2002).
108. Gururaj, A., Barnes, C.J., Vadlamudi, R.K. & Kumar, R. Regulation of phosphoglucomutase 1 phosphorylation and activity by a signaling kinase. *Oncogene* **23**, 8118–8127 (2004).
109. Zhang, S. et al. Rho family GTPases regulate p38 mitogen-activated protein kinase through the downstream mediator PAK1. *J. Biol. Chem.* **270**, 23934–23936 (1995).
110. Brown, J.L. et al. Human Ste20 homologue hPAK1 links GTPases to the JNK MAP kinase pathway. *Curr. Biol.* **6**, 598–605 (1996).

111. Frost, J.A., Xu, S., Hutchison, M.R., Marcus, S. & Cobb, M.H. Actions of Rho family small G proteins and p21-activated protein kinases on mitogen-activated protein kinase family members. *Mol. Cell Biol.* **16**, 3707–3713 (1996).
112. Eblen, S.T., Slack, J.K., Weber, M.J. & Catling, A.D. Rac-PAK signaling stimulates extracellular signal-regulated kinase (ERK) activation by regulating formation of MEK1-ERK complexes. *Mol. Cell Biol.* **22**, 6023–6033 (2002).
113. Sun, H., King, A.J., Diaz, H.B. & Marshall, M.S. Regulation of the protein kinase Raf-1 by oncogenic Ras through phosphatidylinositol 3-kinase, Cdc42/Rac and PAK. *Curr. Biol.* **10**, 281–284 (2000).
114. Schurmann, A. et al. p21-activated kinase 1 phosphorylates the death agonist bad and protects cells from apoptosis. *Mol. Cell Biol.* **20**, 453–461 (2000).
115. Wolf, D. et al. HIV-1 Nef associated PAK and PI3-kinases stimulate Akt-independent Bad-phosphorylation to induce anti-apoptotic signals. *Nat. Med.* **7**, 1217–1224 (2001).
116. Yuan, Z.Q. et al. ArgBP2gamma interacts with Akt and p21-activated kinase-1 and promotes cell survival. *J. Biol. Chem.* **280**, 21483–21490 (2005).
117. Jin, S., Zhuo, Y., Guo, W. & Field, J. p21-activated Kinase 1 (PAK1)-dependent phosphorylation of Raf-1 regulates its mitochondrial localization, phosphorylation of BAD, and Bcl-2 association. *J. Biol. Chem.* **280**, 24698–24705 (2005).
118. Mazumdar, A. & Kumar, R. Estrogen regulation of PAK1 and FKHR pathways in breast cancer cells. *FEBS Lett.* **535**, 6–10 (2003).
119. Vadlamudi, R.K. et al. Regulatable expression of p21-activated kinase-1 promotes anchorage-independent growth and abnormal organization of mitotic spindles in human epithelial breast cancer cells. *J. Biol. Chem.* **275**, 36238–36244 (2000).
120. Bagheri-Yarmand, R. et al. Etk/Bmx tyrosine kinase activates PAK1 and regulates tumorigenicity of breast cancer cells. *J. Biol. Chem.* **276**, 29403–29409 (2001).
121. Faure, S. et al. Control of G2/M transition in Xenopus by a member of the p21-activated kinase (PAK) family: a link between protein kinase A and PAK signaling pathways? *J. Biol. Chem.* **274**, 3573–3579 (1999).
122. Chung, C.Y. & Firtel, R.A. PAKa, a putative PAK family member, is required for cytokinesis and the regulation of the cytoskeleton in Dictyostelium discoideum cells during chemotaxis. *J. Cell Biol.* **147**, 559–576 (1999).
123. Cvrckova, F., De Virgilio, C., Manser, E., Pringle, J.R. & Nasmyth, K. Ste20-like protein kinases are required for normal localization of cell growth and for cytokinesis in budding yeast. *Genes Dev.* **9**, 1817–1830 (1995).
124. Qyang, Y. et al. The p21-activated kinase, Shk1, is required for proper regulation of microtubule dynamics in the fission yeast, Schizosaccharomyces pombe. *Mol. Microbiol.* **44**, 325–334 (2002).
125. Adam, L., Vadlamudi, R., Mandal, M., Chernoff, J. & Kumar, R. Regulation of microfilament reorganization and invasiveness of breast cancer cells by kinase dead p21-activated kinase-1. *J. Biol. Chem.* **275**, 12041–12050 (2000).
126. Nowak, S.J. & Corces, V.G. Phosphorylation of histone H3: a balancing act between chromosome condensation and transcriptional activation. *Trends Genet.* **20**, 214–220 (2004).
127. Li, F. et al. p21-activated kinase 1 interacts with and phosphorylates histone H3 in breast cancer cells. *EMBO Rep.* **3**, 767–773 (2002).
128. Banerjee, M., Worth, D., Prowse, D.M. & Nikolic, M. PAK1 phosphorylation on t212 affects microtubules in cells undergoing mitosis. *Curr. Biol.* **12**, 1233–1239 (2002).
129. Thiel, D.A. et al. Cell cycle-regulated phosphorylation of p21-activated kinase 1. *Curr. Biol.* **12**, 1227–1232 (2002).
130. Muranen, T., Gronholm, M., Renkema, G.H. & Carpen, O. Cell cycle-dependent nucleo-cytoplasmic shuttling of the neurofibromatosis 2 tumour suppressor merlin. *Oncogene* **24**, 1150–1158 (2005).
131. Kohno, M. & Pouyssegur, J. Targeting the ERK signaling pathway in cancer therapy. *Ann. Med.* **38**, 200–211 (2006).

132. Frost, J.A. et al. Stimulation of NFkappa B activity by multiple signaling pathways requires PAK1. *J. Biol. Chem.* **275**, 19693–19699 (2000).
133. Lee, S.R. et al. AR and ER interaction with a p21-activated kinase (PAK6). *Mol. Endocrinol.* **16**, 85–99 (2002).
134. Yang, F. et al. Androgen receptor specifically interacts with a novel p21-activated kinase, PAK6. *J. Biol. Chem.* **276**, 15345–15353 (2001).
135. Michalides, R. et al. Tamoxifen resistance by a conformational arrest of the estrogen receptor alpha after PKA activation in breast cancer. *Cancer Cell* **5**, 597–605 (2004).
136. Rayala, S.K. et al. Functional regulation of oestrogen receptor pathway by the dynein light chain 1. *EMBO Rep.* **6**, 538–544 (2005).
137. Alahari, S.K. Nischarin inhibits Rac induced migration and invasion of epithelial cells by affecting signaling cascades involving PAK. *Exp. Cell Res.* **288**, 415–424 (2003).
138. Huang, Q., Shen, H.M. & Ong, C.N. Emodin inhibits tumor cell migration through suppression of the phosphatidylinositol 3-kinase-Cdc42/Rac1 pathway. *Cell Mol. Life Sci.* **62**, 1167–1175 (2005).

Chapter 6
Molecular Basis for Vascular Endothelial Growth Factor Expression in Tumor

Keping Xie and James Yao

Abstract Vascular endothelial growth factor (VEGF) critically regulates tumor angiogenesis through its potent functions as a stimulator of endothelial cell survival, mitogenesis, migration, and differentiation, as well as vascular permeability, immunosuppression and mobilization of endothelial progenitor cells from the bone marrow into the peripheral circulation. Genetic alterations and a chaotic tumor microenvironment are clearly attributed to numerous abnormalities in the expression and signaling of VEGF and confer a tremendous survival and growth advantage to vascular endothelial cells as manifested by exuberant tumor angiogenesis and a consequent malignant phenotype. Designing effective therapeutic strategies targeting VEGF to control tumor growth and metastasis requires understanding the molecular mechanisms of both inducible and constitutive VEGF expression.

Keywords Vascular Endothelial Growth Factor (VEGF) · Tumor suppressor gene · Hypoxia · Acidosis · Angiogenesis · Neuropilins · Tyrosine kinase · Nitric oxide · Hypoxia-inducible factor-1 (HIF-1) · Promoter analysis · Translational regulation

6.1 Introduction

Vascular endothelial growth factor (VEGF) was initially identified as vascular permeability factor (VPF) by Senger et al. [1] in 1983. They showed that this protein promotes extravasation of proteins from tumor-associated blood vessels. In 1989 two groups independently characterized VEGF as a heparin-binding protein mitogen specific for endothelial cells. Evidently, VPF and VEGF are the same protein encoded by a single gene [2, 3]. Rapidly increasing reports have shown that VEGF is critical in tumor angiogenesis. Interestingly, VEGF has other functions such as antiapoptosis activity, lymphangiogenesis [4, 5], immunosuppression [6, 7], stimulation and recruitment of bone-marrow-derived endothelial and hematopoietic

K. Xie
Department of Gastrointestinal Medical Oncology, The University of Texas, MD Anderson Cancer Center TX 77030, USA
e-mail: kepxie@mail.mdanderson.org

M. Bar-Eli (ed.), *Regulation of Gene Expression in the Tumor Environment*,

precursor cells in angiogenesis [8,9], and regulation of hematopoietic stem cell survival [10]. Therefore, VEGF is indispensable for tumor development and progression and companies are developing treatment modalities targeting VEGF and/or its receptors. However, to help researchers and clinicians more effectively target VEGF and its pathway, it is critical to understqand the molecular mechanisms governing VEGF expression and regulation in tumor microenvironment.

6.2 VEGF Family and Tumor Biology

The VEGF family consists of VEGF or VEGF-A for new classification (6p12-p21) [11], VEGF-B (11q13) [12], VEGF-C (4q34) [13], VEGF-D (Xp22.31) [14], VEGF-E (viral) [15], and an additional member called placenta growth factor (PlGF) (14q24-q31) [16].

6.2.1 VEGF Family Members

The human VEGF gene, which is localized in chromosome 6p12, is organized in eight exons separated by seven introns. The coding region spans approximately 14 kb. The initial discoveries of VEGF were followed by the identification of several splicing variants of VEGF gene transcripts, each encoding an active protein product. At least five VEGF protein products have been identified and designated as VEGF121, VEGF145, VEGF165, VEGF189, and VEGF206. The numbers following "VEGF" indicate the number of amino acids in the respective human VEGF protein. Different isoforms of VEGF proteins have different affinity for heparin and heparin sulfate. Most cell types express multiple variants of VEGF, predominantly VEGF121 and VEGF165.

Structurally, VEGF-B is closely related to VEGF [17]. VEGF-B has two splice variants: the 167-amino-acid form binds to its receptor, while the 186-amino-acid form is a freely secreted, soluble product [18]. Both of the VEGF-B isoforms are able to form heterodimers with VEGF-A and may be able to with other growth factors. VEGF-C was cloned from human prostate carcinoma cells. Its mature form, which consists of the VEGF homology domain, is 30% identical to VEGF165 [13]. VEGF-C is synthesized as a preproprotein, from which a stepwise proteolytic process generates several forms with sequentially increasing binding and activity for its receptors. VEGF-D (also known as c-*fos* – induced growth factor) is the most recently discovered member of the mammalian VEGF family [19]. It shares 61% sequence identity with and is proteolytically processed similarly to VEGF-C. The proteolytic processing also appears to regulate the biological activity and receptor specificity of VEGF-D [20]. VEGF-E was discovered in the genome of *Parapoxvirus*, orf virus. Two forms of VEGF-like molecules have been identified – OV-VEGF2 and OV-VEGF7 – which are most closely related in primary structure to VEGF, sharing 29% and 23% amino acid identity, respectively, with human VEGF121 [21]. PlGF, which was discovered in the human placenta, is around 50% homologous with VEGF [16]. Three splice isoforms of PlGF have been identified, and one of them, PlGF2, at least competes with VEGF165 for binding to their receptor [22, 23]. PlGF/VEGF heterodimers are more potent in mediating biological activity than PlGF homodimers are [24]. All of the proteins in the VEGF family have a similar molecular structure, but they show distinguishable spectra of functions as well as binding specificity toward their shared receptors.

6.2.2 VEGF Receptors

VEGF family members signal by binding to members of a group of at least four high-affinity receptors. Three of these receptors are cell surface proteins of the receptor tyrosine kinase family: VEGFR-1 (*fms*-like tyrosine kinase-1), VEGFR-2 (kinase insert domain-containing receptor, fetal liver kinase), and VEGFR-3 (FLT-4). These three receptors have seven extracellular immunoglobulin homology domains and two intracellular tyrosine kinase domains split by a kinase insert and ending with a *C*-terminal tail within their cytoplasmic region [25]. The ligand specificities of these receptors are as follows: VEGFR-1 binds VEGF-A, VEGF-B, PlGF1, and PlGF2; VEGFR-2 binds VEGF-A, -C, -D, and -E; and VEGFR-3 binds VEGF-C and -D [26] VEGFR-1 and -2 are selectively though not exclusively expressed on vascular endothelium, whereas VEGFR-3 is expressed mainly on lymphatic endothelium. Nonetheless, it is apparent that VEGFR-2 is also present on lymphatic endothelium and that VEGFR-3 can be expressed on the endothelium of blood vessels [27]. Neuropilin-1 and -2 (NRP-1 and -2) constitute another class of high-affinity nontyrosine kinase receptors of VEGFs and can bind certain isoforms of VEGF-A, -B, and -E and PlGF-2. Neuropilins have a short cytoplasmic tail with no known signaling function, so by themselves they may not be functional receptors. NRP-1 was described as a co-receptor specific for VEGF165, because it increases the affinity of VEGF165 to VEGFR-2 by about 10-fold, resulting in VEGF165 being the strongest signal transducer among the VEGF subtypes. On the other hand, no responses to VEGF165 have been observed following VEGF165 stimulation of cells expressing NRP-1 but no other VEGF receptors [26]. NRP-1 is expressed on vascular endothelium, neurons, and some tumor cells [26] and is co-expressed with VEGFR-1 and VEGFR-2 [28]. Recently, NRP-2 has been shown to bind VEGF-C and be co-expressed with VEGFR-3 in the endothelial cells of certain lymphatic vessels [29].

The active forms of VEGF-A are mostly homodimers with a molecular weight of 45 kDa [30]. Heterodimers formed by PlGF and VEGF-A have been detected in certain cells and function as endothelial mitogens [24]. Also, VEGF-A and VEGF-B heterodimers have been obtained in an expression vector co-transfection experiment [31]. The two monomers of VEGF-A are assembled in an antiparallel fashion to form the homodimer, which in turn binds to domains 2 and 3 of the seven immunoglobulin-like domains in the receptors. Upon binding to its receptor, VEGF initiates a cascade of signaling events that begins with dimerization of two receptors and then autophosphorylation of each other by the tyrosine kinase domain to form the active receptors. This is followed by activation of numerous downstream proteins, including phospholipase C-γ/protein kinase C (PKC), *Ras* pathway members, phosphatidylinositol-3 kinase (PI3K), mitogen-activated protein kinase (MAPK), and others [see review [25, 28, 32–34]] to manifest end point function, such as an increase in vascular permeability, cell survival and proliferation, and migration.

6.2.3 Biological Functions of VEGF

It is well recognized that VEGF plays a critical role in many aspects of cancer biology. Specifically, VEGF is mitogenic, motogenic, and morphogenic in endothelial cells and key to tumor angiogenesis and lymphangiogenesis, which are indispensable for tumor development and progression. Recent studies have indicated that VEGF appears to play a broader role in cancer pathogenesis than previously thought. For example, a fundamental cellular mechanism by which VEGF promotes the formation of new blood vessels and maintains their integrity is the activation of endothelial cell survival and antiapoptotic pathways [4,5]. Overproduction of VEGF may also be related to tumor-associated immunosuppression [6,7,35]. A direct association between increased levels of VEGF expression in tumor cells and low number of antigen-presenting dendritic cells in the vicinity of tumors was established in 140 patients with gastric cancer [36]. Continuous infusion of VEGF resulted in dramatic inhibition of dendritic cell development by diminishing the ability of hematopoietic progenitor cells to differentiate into functional dendritic cells during the early stages of their maturation [37, 38]. Also, continual administration of recombinant VEGF in naïve mice resulted in inhibition of dendritic cell development and increased production of B cells and immature Gr-1$^+$ myeloid cells [38]. Neutralizing the anti-VEGF antibody blocked the negative effects of tumor cell supernatants on dendritic cell maturation in vitro [39] and significantly improved the number and function of lymph node and spleen dendritic cells in tumor-bearing animals, thereby enhancing the efficacy of cancer immunotherapy [40]. In addition, a recent report showed that VEGF can inhibit the development of T cells from early hematopoietic progenitor cells, which may be another mechanism contributing to tumor-induced immunosuppression [6]. On the other hand, VEGF clearly induces stimulation and recruitment of bone-marrow-derived endothelial and hematopoietic precursor cells in angiogenesis [8, 9] and regulates hematopoietic stem cell survival [10]. VEGF also has many secondary effects via induction of a number of active substances that have a wide range of actions, including nitric oxide (NO), plasminogen activators, and endothelial cell decay-accelerating factor [41–45].

6.2.4 Clinical Significance of VEGF

The demonstrated role of VEGF in tumor biology makes its clinically significant in cancer diagnosis, prognosis, and treatment. Studies have shown that the VEGF expression level is significantly higher in tumors than in normal tissue and is correlated with tumor grade, depth of invasion, status of nodal and distant metastasis, and TNM and clinical stage [46–51]. VEGF is also an important prognostic factor in many human tumors [52,53]. The tissue expression level of VEGF has been correlated with patient survival in breast cancer [54], ovarian cancer [55], fibrillary low-grade astrocytoma [56], colorectal cancer [57], non-small cell lung carcinoma (NSCLC) [58], and pancreatic cancer [59]. Also, serum VEGF levels have been shown to be correlated with survival in patients with ovarian cancer [60], gastrointestinal cancer [61,62], or acute myeloid leukemia [63,64]. Due to the critical role of VEGF in tumor growth and metastasis, targeting VEGF and VEGF receptor signaling in cancer treatment has been attempted with success using many approaches for various tumors. Clinical trials testing the efficacy of many angiogenesis inhibitors that are based on targeting VEGF and its signal transduction pathways are under way [65,66].

6.3 Regulation of VEGF Expression

Regulation of VEGF expression has been reported to occur at the gene transcription, translation, and posttranslation levels. Transcriptional regulation of VEGF expression has been studied extensively, because the impact of most genetic and epigenetic factors on VEGF expression is realized by controlling VEGF gene transcription. Computer-based sequence analysis of the VEGF gene promoter structure revealed a number of potential binding sites in the 5'-flanking region of the VEGF gene for specific protein-1 (Sp1), hypoxia-inducible factor 1 (HIF-1), signal transducer and activator of transcription-3 (Stat3), activator protein-1 (AP-l), Egr-1, activator protein-2 (AP-2), nuclear factor-IL6, and many others [67–69], indicating the diverse complexity of VEGF transcriptional regulation. Among the many transcription factors, Sp1, HIF-1, Stat3, and AP-1 appear to be the key factors in regulation of VEGF expression and have been well characterized.

6.3.1 Transcriptional Regulation

Sp1. Sp1 was the first eukaryotic transcription factor to be identified and cloned [75] and has been shown to stimulate transcription through binding to G/C-rich boxes present on a wide variety of promoters. Sp1 is a highly phosphorylated protein. Phosphorylation modification can regulate the transcriptional activity of Sp1 by affecting its DNA binding ability [76]. Detailed promoter analysis using reporter gene assays, electrophoretic mobility shift assays, and mutagenesis of promoter elements has been applied in our laboratory to characterization of the *cis*-responsive elements in the VEGF promoter, resulting in the identification of four G/C-rich putative Sp1 binding sites at the region −38 to −109 bp relative to the single transcriptional start site. Further deletion and point mutation analyses indicated that mutation of some or all of the putative Sp1 binding sites reduced or eliminated the constitutive VEGF promoter activity and abrogated the differential activity of the promoter in high and low VEGF-expressing cells [70]. Similarly, Ryuto et al. [77] found that the four Sp1 binding sites are essential for basal transcription of the VEGF gene and tumor necrosis factor-α (TNF-α)-dependent promoter activation in a human glioma cell line. Deletion of these Spl binding sites reduced basal transcription of VEGF and abolished TNF-α responsiveness. Furthermore, many genetic alterations affect VEGF expression through modulation of the transcriptional activity of Sp1. These include alterations of tumor suppressor genes, such as von Hippel-Lindau (VHL) [78–80], p53 [81,82], and p73 [83], and oncogenes, such as *Ras*, *Src*, and *HER2/neu* [84,85]. In addition, VEGF expression can be regulated through modulation of Sp1 activity by tumor microenvironmental factors, such as free radicals, hypoxia, and growth factors [77,86,87].

HIF-l. HIF-1 is a transcriptional activator composed of HIF-1α and HIF-1β (also called aryl hydrocarbon receptor nuclear translocator) subunits. There are several dozen known HIF-1 targets, including the VEGF gene. Both HIF-1α and HIF-1β are constitutively expressed in various types of tumor. Under normal oxygenation conditions, HIF-1α is barely detectable, because it is rapidly degraded by the ubiquitin-proteasome system, with a half-life of only 5 minutes. This degradation is controlled by the oxygen-dependent degradation domain within HIF-1α. Under hypoxic conditions, HIF-1α expression increases as a result of decreased ubiquitination and degradation. The tumor suppressors VHL and p53 target HIF-1α for ubiquitination such that inactivation of them in tumor cells increases the half-life of HIF-1α. Increased PI3K and Akt and decreased phosphatase and tensin homologue deleted on chromosome 10 (PTEN) activity in prostate cancer cells also increase HIF-1α expression by an undefined mechanism. HIF-1 activates transcription of the VEGF gene by binding to the hypoxic response element (HRE) in the gene promoter. A HIF-l binding site was identified at −975 to −968 bp of the human VEGF promoter [88]. Recent studies revealed that alteration of both tumor suppressor genes and oncogenes constitutively increases the transcriptional activity of HIF-1 via three different molecular mechanisms, which include those affecting the stability, expression, and transcriptional activity of HIF-1. These genetic alterations include loss of VHL, p53, or PTEN function and gain-of-function mutations in the Ras and v-Src oncogenes. In addition to genetic alterations, the protein stability and transcriptional activity of HIF-1 are also modulated by tumor microenvironmental factors, such as hypoxia, growth factors, cytokines, and oxidative stress [89].

Stat3. Lately, we and others have found that the transcription factor Stat3, a member of the JAK-STAT signaling pathway, plays an important role in the regulation of VEGF expression in tumors [67,73,90]. This is supported by a significant amount of evidence. Activation of Stat3 has been observed in cells transformed in vitro with the v-*Src* and *Abl* oncogenes. Constitutive activated Stat3 protein is found in various types of tumors, with activation of it correlated with the level of VEGF expression. In fact, Stat3 can directly bind to the VEGF promoter in vivo and upregulate VEGF promoter activity and protein levels. Mutagenesis of putative Stat3 binding sites, which are located at −842 to −849 bp, significantly reduces the VEGF promoter activity induced by activated Stat3. On the other hand, blockage of activated Stat3 via ectopic expression of dominant-negative Stat3 significantly suppresses VEGF promoter activity, VEGF expression, angiogenesis, and tumor growth and metastasis. These data indicate that VEGF is directly regulated by Stat3 [73, 91, 92]. Moreover, Stat3 is known to be activated by numerous cytokines and growth factors, including epidermal growth factor (EGF), platelet-derived growth factor (PDGF), VEGF and interleukin-6, suggesting that Stat3 signaling may be a common molecular target for blocking angiogenesis in human tumors [90, 93–96].

AP-1. AP-1 is a transcription factor belonging to the leucine zipper family. This factor is a dimer composed of jun/jun (c-jun, junB, and junD) or jun/fos subunits (s-fos, fosB, fra-1, and fra-2). In VEGF promoter, there are four AP-1 candidate binding sites [67], indicating that AP-1 might be involved in the regulation of VEGF gene expression. The signaling for AP-1 activation was reported to occur through the protein kinase C (PKC) pathway [97–99] and the MAP kinase pathway [100, 101]. For example, PKC inhibitors suppress VEGF induction in glioma cells by EGF, PDGF, bFGF [102] Hypoxia, oxidative stress, ultraviolet irradiation, and cytokines may induce VEGF expression through the synthesis of jun and fos proteins, leading to increased AP-1 binding activity [101, 103–105].

Many other transcription factors also contribute to VEGF expression and regulation, such as AP-2 [106, 107], Smad3 and Smad4/DPC4 [108, 109]. Moreover, optimal transcriptional activation of VEGF may require the activation and cooperation of different transcription factors and/or signaling pathways. For example, VEGF induction by hypoxia or NO requires the involvement of HIF-1 and is potentiated by AP-1 [72, 110]. Hypoxia and TGF-β can synergize in the regulation of VEGF expression at the transcriptional level, possibly through physical interaction and functional cooperation between Smads and the HIF-1α transcription factor [71]. Additionally, the optimal transactivation of VEGF by p42/p44 MAPK signaling pathway needs the cooperation of Sp1 and AP-2 [111] and full induction of VEGF expression by ultraviolet B radiation requires the cooperation of Sp1 with AP-1 [112].

6.3.2 Posttranscriptional Regulation

Even though transcriptional regulation represents the most important mechanism of VEGF expression and regulation, posttranscriptional regulation also plays an important part in VEGF expression. Levy et al. [113] first reported the stabilization of VEGF mRNA by hypoxia and identified a region in the VEGF 3'-untranslated region (UTR) responsible for the stabilization. Subsequently, the protein responsible for the stabilization was identified as HuR, an AU-rich element binding protein [114]. Also, Nabors et al. [115] investigated the pattern of expression of HuR in 35 freshly resected and cultured central nervous system tumors. They found that HuR mRNA was consistently expressed in all of the tumors regardless of the cell origin or degree of malignancy. However, using a novel HuR-specific polyclonal antibody, they found that strong HuR protein expression was limited to high-grade malignancies (glioblastoma multiforme and medulloblastoma). Within glioblastoma multiforme, prominent HuR expression also was detected in perinecrotic areas (hypoxic), in which angiogenic growth factors are upregulated. An enzyme-linked immunosorbent assay-based RNA binding assay showed that HuR specifically binds to the AU-rich stability elements located in the 3'-UTR of VEGF mRNA. Recently, Dibbens et al. [116] demonstrated that three elements located in the 5'-UTR and 3'-UTR were all required for maximum stabilization of the mRNA during hypoxia. These findings suggest a role for HuR protein in the posttranscriptional regulation of VEGF expression in tumors [115]. Although the detailed signaling pathway remains unclear, the stress-activated protein kinases c-jun N-terminal kinase (JNK) and p38/HOG appear to be indispensable for the stabilization of VEGF mRNA [117].

6.3.3 Translational Regulation

It has also been proposed that VEGF expression can be regulated at the translational level [118]. Among a series of control mechanisms exerted at the translational level, the use of alternative codons is a very subtle means of increasing gene diversity by expressing several proteins from a single mRNA. At least five isoforms of VEGF-A are generated this way. VEGF isoforms have distinct activities at different anatomical sites, and the microenvironment of different tissues affects the expression and function of VEGF isoforms [119]. Different VEGF isoforms have distinct activities at different anatomical sites, and the microenvironment of different tissues affects the expression and function of VEGF isoforms [119]. For example, the internal ribosome entry sites act as specific translational enhancers that allow translation initiation to occur independently of the classic cap-dependent mechanism in response to specific stimuli and under the control of different trans-acting factors. Also, the 5'-UTR of VEGF mRNA contains two functional internal ribosome entry sites that maintain efficient cap-independent translation and ensure efficient production of VEGF even under unfavorable stress conditions, such as hypoxia, which globally decrease the rate of translation initiation [120, 121]. Although much progress has been made in unraveling the diverse complexity of the molecular regulation of VEGF expression, our understanding of this process continues to grow. For example, several other factors are important to the translational regulation of VEGF expression. Kevil et al. [122] revealed that enforced overexpression of eukaryotic initiation factor 4E drastically increases VEGF secretion (up to 130-fold) in Chinese hamster ovary cells. Eukaryotic initiation factor 4E is a 25-kDa polypeptide that recruits mRNAs for translation by binding to the 7-methylguanosine – containing cap of mRNA. Overexpression of it has been observed in metastatic breast carcinomas [123] and bladder cancer [124], suggesting a possible contribution to tumor growth and progression. In addition, ORP150 is a 150-kDa oxygen-regulated protein whose expression is induced by hypoxia [125]. Physiologically, ORP150 functions as a molecular chaperone in the endoplasmic reticulum for the folding and trafficking of newly synthesized protein to the Golgi apparatus for subsequent secretion. Overexpression of ORP150 promotes VEGF protein secretion into hypoxic culture supernatants, whereas expression of ORP150 antisense RNA results in accumulation of VEGF within the endoplasmic reticulum [126]. Administration of an adenovirus encoding ORP150 to wounds in diabetic mice accelerated neovascularization and wound repair in vivo Immunohistochemical analysis showed that expression of ORP150 and VEGF was enhanced in the cytoplasm of prostate cancer cells. Additionally, adenovirus-mediated transduction of antisense ORP150 into DU145 human prostate cancer cells resulted in decreased VEGF production, angiogenesis, and marked suppression of tumor formation in the xenograft animal model [127]. These data indicate that OPR150 is required for VEGF secretion and that, in hypoxic cells, increased levels of VEGF expression necessitate a corresponding increase in ORP150 expression.

6.4 Signaling Pathways for Constitutive VEGF Expression

Autonomous growth is a key characteristic of a malignant tumor. Usually, the tumor cells can supply themselves (autocrine secretion) or manipulate host stromal cells to supply them (paracrine secretion) with what they need to grow. Numerous reports have shown that tumor growth and progression are closely related to VEGF expression. Specifically, it has been reported that many cancer cells constitutively express VEGF proteins without apparent stimuli, which may provide a paracrine mechanism to induce angiogenesis and/or an autocrine mechanism to induce proliferation if the cells also express VEGF receptors. While little is currently known about the molecular regulation leading to constitutive VEGF expression, it is apparent that the genetic makeup of tumor cells is most probably involved in constitutive VEGF expression by affecting the production and function of transcriptional regulators [128]. Among the many important genetic alterations, mutations of various oncogenes and tumor suppressor genes can profoundly affect the downstream signal transduction pathways critical to VEGF expression. For example, 70% of human pancreatic adenocarcinoma cell lines constitutively overexpress VEGF. Additionally, the constitutive levels of Sp1 activity are directly correlated with the constitutive levels of VEGF expression [70]. It has been demonstrated that wild-type tumor suppressors such as p53, p73, VHL can physically interact with Sp1, form a complex, and block Sp1 binding to the promoter of VEGF, thus inhibiting Sp1-mediated VEGF expression. Recent studies have consistently shown that loss or inactivation of the wild-type VHL, p53, and/or p73 gene is associated with increased tumor angiogenesis [83, 129].

6.4.1 Activation of Oncogenes

Several oncogenes have been implicated in increased VEGF expression, including activated forms of *Ras*, *Src*, *HER2/neu*, and *Bcr/Abl* [84, 85]. However, mutant oncogene-dependent VEGF expression is necessary, but not sufficient, for progressive tumor growth, suggesting a relative contribution of oncogenes, such as mutant K-*Ras*, to tumor angiogenesis [130].

Ras. Expression of a mutant *Ras* oncogene is one of the most commonly encountered genetic changes detected in human cancer. Rak et al. [131] first provided direct evidence that mutant *Ras* oncogenes upregulate VEGF/VPF expression and tumor angiogenesis, which was immediately confirmed by many investigators in studies of various types of human and animal tumors [132–134]. In general, elevated levels of both VEGF mRNA and secreted functional protein expression can be detected in human and rodent tumor cell lines expressing mutant K-*Ras* and H-*Ras* oncogenes, respectively. Genetic disruption of the mutant K-*Ras* allele in human colon carcinoma cells has been associated with a reduction in VEGF activity. Furthermore, pharmacological disruption of mutant Ras protein function in H-*Ras*-transformed rat intestinal epithelial cells via treatment with L-739, 749 (a protein farnesyltransferase inhibitor) has been shown to cause significant suppression of VEGF activity.

The signaling pathways of *Ras*-mediated VEGF regulation are not entirely clear. Like several other tyrosine kinase oncogenes, activated *Ras* has been shown to stabilize VEGF mRNA [134]. Most notably, activated *Ras* activates downstream kinases through two distinct pathways – the MAPK pathway (Raf/MEK/p42/p44 MAPK) and the PI3K/Akt pathway – thereby causing the phosphorylation and activation of both MAPK and Akt [135], which lead to enhanced VEGF gene transcription.

Pouysségur et al. [136] recently showed that the MAPK pathway plays a critical role in transcription of VEGF gene regulation, one of the important downstream targets of which is Sp1. In fact, p42/p44 MAPK directly phosphorylates Sp1 on threonine 453 and 739, increases Sp1 DNA binding activity, and thus upregulates VEGF transcriptional expression. They also found that a G/C-rich region of the VEGF promoter from −88 to −66 bp that contains two Sp1 binding sites and one AP-2 binding site is responsible for the upregulation of VEGF promoter activity by p42/p44 MAPK activation. The maximal transcriptional activation requires the cooperation of Sp1 and AP-2, because individual mutations of one AP-2 and two Sp1 putative binding sites do not significantly modify the basal and constitutively activated form of MAPK kinase kinase (MKK1SS/DD)-stimulated VEGF promoter activity, but combined mutation of the AP-2 binding site and both Sp1 binding sites dramatically decreases the basal and MKK1SS/DD-dependent transcriptional activation of the VEGF promoter [111]. In addition, a new pathway for Sp1-mediated VEGF transcriptional activation by a *Ras* signaling pathway has been revealed through the identification of PKC-ζ as the downstream effector of both MAPK and PI3K/Akt [76, 137]. In fact, PKC-ζ can directly interact with Sp1 and phosphorylate the zinc finger region of Sp1, which increases Sp1 DNA binding activity, resulting in Sp1-mediated transcriptional activation [138]. This finding is also the first demonstration of a direct connection between the PI3K/Akt pathway and Sp1 transcriptional activation. Constitutive activation of the MAPK and PI3K/Akt pathways has been observed in many tumors [139–141] and plays a very important role in tumor progression [142–144].

On the other hand, Mazure et al. [145] showed that *Ras*-transformed cells do not use the downstream effectors c-Raf-1 and MAPK in signaling VEGF induction by hypoxia, as overexpression of kinase-defective alleles of these genes does not inhibit VEGF induction under low oxygen conditions. In contrast with the c-Raf-1/MAPK pathway, hypoxia increases PI3K activity in a *Ras*-dependent manner, and inhibition of PI3K activity genetically and pharmacologically results in inhibition of VEGF induction. Therefore, hypoxia modulates VEGF induction in *Ras*-transformed cells through activation of a stress-inducible PI3/Akt pathway and the HIF-1 transcriptional response element. In addition, activation of p42/p44 MAPK induces HIF-1α phosphorylation and promotes HIF-1 – dependent transcriptional activity, resulting in upregulation of VEGF expression [146]. These results suggest a novel mechanism whereby two divergent signaling pathways emerging from *Ras* may cooperatively but independently regulate the activity of a HIF-1α, thereby promoting VEGF expression [135].

The regulatory roles of *Ras* in VEGF expression may be more complex than the use of p42/p44 MAPK and PI3/Akt pathways. It has been suggested that oxidative stress induction of VEGF is dependent on activation of AP-1, which is involved in the activity of Ras/Raf-1/MEK/ERK1/2 [147]. The *Ras* pathway can also be influenced by or interact with many other signaling pathways. For example, extracellular signal-regulated kinase, p38 kinase, and Eph kinases differentially regulate the *Ras*-mediated stimulation of VEGF [148, 149]. In addition, the mutated Wnt signaling pathway has been found to strongly upregulate VEGF, and K-*Ras* activation appears to enhance Wnt signaling, suggesting a unique interaction between these two pathways [150]. Furthermore, new evidence indicates that cell contact induction of VEGF transcription requires the activity of focal adhesion kinase, Rap1, and Raf, which represents a *Ras*-independent mechanism [151].

Src. *Src* was initially discovered as the oncogenic protein (v-Src) of the retrovirus Rous sarcoma virus; a ubiquitously expressed and highly conserved cellular homologue of v-Src was subsequently discovered. Much of the original research on v-Src and its cellular homologue (Src) paved the way for our current understanding of how oncogenes cause cell transformation. The Src family of nonreceptor protein tyrosine kinases, which also includes Lyn, Fyn, Lck, Hck, Fgr, Blk, and Yes, has important roles in receptor signaling and cellular communication. While most Src kinases are broadly expressed (e.g., Src, Fyn, and Yes), certain members of the Src family, such as Hck, Blk, and Lck, exhibit restricted tissue expression [152]. Recent studies have demonstrated that, in addition to their central role in receptor signaling and cell communication, Src kinases play an important role in lymphokine-mediated cell survival and VEGF-induced angiogenesis. Src kinases are activated by a variety of cell surface receptors.

Mukhopadhyay et al. [84] were the first to show that genistein, an inhibitor of protein tyrosine kinase, blocks VEGF induction. Hypoxia increases the kinase activity of pp60c-src and its phosphorylation on tyrosine 416 but does not activate Fyn or Yes. Expression of either a dominant-negative mutant form of c-Src or Raf-1 markedly reduces VEGF induction. VEGF induction by hypoxia in c-Src(-) cells is impaired, although this impairment causes compensatory activation of Fyn, suggesting that VEGF is a new downstream target for c-Src. This finding was confirmed using a variety of protein tyrosine kinase inhibitors [153, 154]. Overexpression of v-Src upregulates VEGF expression by activating a VEGF promoter-luciferase construct in a dose-dependent manner, which is opposed by the presence of wild-type p53 [155]. In fact, p53 can inhibit VPF/VEGF expression by downregulating Src kinase activity under both normoxic and hypoxic conditions [82]. When the steady-state level of pp60c-src is reduced in HT-29 colon adenocarcinoma cells transfected with a c-Src antisense expression vector, not only is the steady-state level of VEGF reduced, but the ability of confluence to stimulate pp60c-src activity and VEGF production is, too. These data suggest that c-Src may be an intermediary in both constitutive and inducible pathways for VEGF production in colon tumor cells [156]. Northern blot analysis of such cell lines revealed that VEGF mRNA expression was decreased in proportion to the decrease in Src kinase activity. Under hypoxic conditions, cells with decreased Src activity had less than a twofold increase in VEGF expression, whereas parental cells had greater than a 50-fold increase. VEGF expression in the supernatants of cells was also reduced in antisense transfectants compared with that in parental cells. In nude mice, subcutaneous tumors obtained from antisense transfectants showed a significant reduction in vascularity. These results suggest that Src activity regulates the expression of VEGF in colon tumor cells [157].

However, there is evidence against a regulatory role for Src kinase. For example, in Hep3B cells, transient, stable transfection substantially changed Src activity, but no alteration was seen in the normoxic or hypoxic expression of erythropoietin, VEGF, or Glut-1 or in the regulation of hypoxia-inducible HIF-1 – dependent reporter genes. Similarly, the expression of these genes in Src- and c-Src – kinase-deficient cells did not differ from that in wild-type cells at either 1% oxygen or more severe hypoxia. These results indicate that Src is not critical for the hypoxic induction of HIF-1, erythropoietin, VEGF, or Glut-1 expression. Also, in a transgenic mouse model, VEGF and v-Src expression patterns were not identical, suggesting that VEGF activation was not dependent only on v-Src [158]. Jiang et al. [159] demonstrated that while c-Src expression is not required for expression of HIF-1 or transcriptional activation of genes encoding VEGF and enolase 1 (ENO1), cells expressing the v-*Src* oncogene have increased expression of HIF-1, VEGF, and ENO1 under both hypoxic and nonhypoxic conditions. Furthermore, expression of v-*Src* was associated with increased transcription of reporter genes containing *cis*-acting hypoxia-response elements from the VEGF and ENO1 genes; this transcriptional activation required an intact HIF-1 – binding site. Most recently, Lee et al. [160] reported that the Src-suppressed C-kinase substrate decreases VEGF expression through AP-1 reduction.

HER2/neu. c-erbB2 encodes the human epidermal growth factor receptor 2 (HER2, HER-2/neu, or ErbB2/neu), which is a proto-oncogenic receptor tyrosine kinase that is overexpressed, amplified, or both in several human malignancies, including breast, ovarian, and colon cancer [161]. Its regulatory role in VEGF expression was first examined in NIH 3T3 fibroblasts transformed with mutant ErbB2/neu, which resulted in significant induction of VEGF expression. Moreover, treatment of ErbB2/neu-positive SKBR-3 human breast cancer cells in vitro with a specific neutralizing anti-ErbB2/neu monoclonal antibody resulted in a dose-dependent reduction of VEGF protein expression [162]. Direct correlation between the expression levels of VEGF and HER2/neu was then established in human tumor specimens, including those of breast, cervical, head and neck, ovarian, and lung cancer [163–167].

Recently reported data have provided a novel molecular basis for induction of VEGF and tumor angiogenesis by heregulin-HER2 signaling. HER2/neu signaling induced by overexpression of in mouse 3T3 cells and heregulin stimulation in human MCF-7 breast cancer cells result in increased HIF-1α protein and VEGF mRNA expression that is dependent on the activity of PI3K, AKT, and the downstream kinase FRAP(FKBP12-rapamycin-associated protein; also known as mammalian target of rapamycin, mTOR). In contrast with other inducers of HIF-1 expression, heregulin stimulation does not affect the half-life of HIF-1α but instead stimulates HIF-1α synthesis in a rapamycin-dependent manner. The 5'-UTR of HIF-1α mRNA directs heregulin-inducible expression of a heterologous protein. These results suggested that heregulin-HER2/neu signaling mediates VEGF induction via regulation of HIF-1α expression [88]. However, other report showed that Sp1 and AP-2 binding sites within the proximal region of VEGF promoter are required for up-regulation of VEGF by heregulin-β1 and that this up-regulation is dependent on the activity of extracellular signal-related protein kinases [168].

Other oncogenes. Many other oncoproteins besides Ras, Src, and HER2/neu may influence VEGF expression in tumor cells. For example, in chronic myelogenous leukemia, BCR/ABL expression is associated with increased expression of VEGF and HIF-1 [169]. Studies have shown that transfection of BCR/ABL results in enhanced VEGF expression and that treatment with STI-571 (imatinib mesylate, a Bcr/Abl – targeting drug) reduces the level of VEGF expression in BCR/ABL-positive cells [170]. In addition, transactivation of the VEGF promoter by BCR/ABL appears to involve the PI3K/Akt pathway [169]. HPV-16 E6-positive cells generally have a high level of expression of the VEGF message. Furthermore, co-expression of the VEGF promoter reporter gene with E6 in both human keratinocytes and mouse fibroblasts has shown that the E6 oncoprotein upregulates VEGF promoter activity in a p53-independent manner. An E6-responsive region that comprises four Sp1 sites between from -194 to -50 bp of the VEGF promoter is also necessary for constitutive VEGF transcription [171]. Finally, an inverse association was found between bcl2 expression and VEGF activity via immunohistochemistry analysis in NSCLC [172] and hepatocellular carcinoma [163, 173]. Human breast cancer cells transfected with Bcl-2 have exhibited increased levels of VEGF expression in tissue culture and xenografts [174]. Thus, many oncoproteins may contribute to tumor angiogenesis via direct stimulation of the VEGF gene and detailed signaling. However, this remains to be elucidated.

6.4.2 Inactivation of Tumor Suppressor Genes

Tumor suppressor genes have also been implicated in the regulation of VEGF gene expression. These genes include VHL, TP53, p73, and PTEN, and p16.

VHL. Mutation or loss of both alleles of the VHL gene was initially documented in sporadic renal cell carcinomas, central nervous system hemangioblastomas, and neoplasms that arise in individuals having VHL syndrome. In particular, Wizigmann-Voos et al. [175] described the upregulation of VEGF and its receptors in VHL-syndrome-associated and sporadic hemangioblastomas. Deregulated VEGF expression in human renal carcinoma cells was reversed by introduction of the wild-type VHL tumor suppressor protein [176].

Although VHL protein (pVHL) function remains unclear, VHL does interact with the elongin BC subunits in vivo and regulate RNA polymerase II elongation activity in vitro by inhibiting formation of the elongin ABC complex [177]. Gnarra et al. [178] provided evidence that despite the differences in VEGF mRNA levels, VHL overexpression does not affect VEGF transcription initiation or elongation as suggested by the VHL-elongin association. Instead, VHL regulates VEGF expression at a posttranscriptional level.

Iliopoulos et al. [179] first suggested that VHL plays a critical role in the transduction of signals generated by changes in ambient oxygen tension, such as negative regulation of hypoxia-inducible VEGF expression by the pVHL. Consistent with posttranscriptional regulation, a 500-bp region of the 3'-UTR of VEGF mRNA was identified, which is critical for stabilization of VEGF mRNA through the formation of an RNA-protein complex in a hypoxia-inducible fashion. In fact, three adenylate-uridylate-rich RNA elements within this region form an identical or closely related hypoxia-inducible RNA-protein complex. This complex is constitutively elevated in tumor cell lines lacking the wild-type VHL gene and in which VEGF mRNA expression is constitutively stabilized [180].

On the other hand, VHL also regulates VEGF expression at the transcriptional level. Mukhopadhyay et al. [79] described a new mechanism for VHL-mediated transcriptional repression of VEGF expression in which the VHL gene product directly interacts with Sp1 to repress VEGF promoter activity. The VHL Sp1-binding domain has been mapped to amino acids 96-122; this domain is disproportionately affected by substitution mutations, which interfere with the VHL-Sp1 interaction. Deletion of the domain prevents VHL effects on Sp1 DNA binding and VHL target gene expression, indicating that the domain contributes significantly to the tumor suppressor activity of VHL [80]. Also, the VHL gene product inhibits VEGF expression in renal cell carcinoma cells by blocking PKC pathways. Specifically, the wild-type VHL gene neutralizes PKC isoforms ζ and δ by forming cytoplasmic complexes with them. This inhibits MAPK activation, thereby preventing the translocation of these isoforms to the cell membrane, where they otherwise would engage in signaling steps leading to aberrant VEGF overexpression [181].

Recently reported data indicate that VHL can also repress VEGF expression by interacting with HIF-1. The pVHL is a component of an E3 ubiquitin ligase that targets HIF-1α for ubiquitylation and degradation under normoxic conditions. pVHL also directly inhibits HIF-1α transactivation by recruiting histone deacetylases. Thus, VHL mutations can lead to increased transcription of hypoxia-inducible genes such as VEGF [182]. The increased VEGF mRNA stability in cells lacking pVHL has been hypothesized to be due to similar regulation of an RNA-binding protein, the expression of which is regulated by pVHL in a manner that is dependent on elongin C interactions as well as functioning proteasomes [182].

Additionally, a novel pVHL-interacting protein has been identified as a negative regulator of HIF-1α transactivation. This protein, which is named pVHL-associated KRAB-A domain-containing protein (VHLaK) and is generated from the ZnF197 locus via alternative splicing, contains a Kruppel-associated box (KRAB)-A domain and SCAN domain but lacks the 22 C2H2-type zinc fingers present in ZnF197. The KRAB-A domain in VHLaK mediates pVHL binding and functions as a transcriptional repression module. Also, the SCAN domain mediates VHLaK homo-oligomerization, which enhances the repressive activity of VHLaK. pVHL can recruit VHLaK to repress the transcriptional activity of HIF-1α and HIF-1α–induced VEGF expression. Finally, pVHL, VHLaK, and KAP1/TIF-1β can be recruited into a single complex, indicating that KAP1/TIF-1β may participate in pVHL-mediated transcriptional repression of HIF-1α. These findings reveal a novel mechanism for the modulation of HIF-1α transactivation by pVHL [183].

TP53. TP53 is mutated in more than 50% of all human tumors. It exerts its protective effects in part by inducing cell cycle arrest or apoptosis in response to various types of genotoxic stress. TP53 is a sequence-specific DNA binding transcription factor, and some of its gene targets include cell cycle regulators like p21 and GADD45 as well as apoptosis mediators like bax and DR5/KILLER. TP53 has also been linked with tumor angiogenesis. Initially, Kieser et al. [184] showed in transient transfection assays that a mutated form of the murine p53 gene (ala135⟶val) induces expression of VEGF mRNA and potentiates (TPA)-stimulated VEGF mRNA expression. In NIH 3T3 cells that stably overexpress the temperature-sensitive p53 (ala135⟶val) and display the mutant phenotype at 37°C and wild-type phenotype at 32.5°C, induction of VEGF mRNA and protein expression by activated PKC is strongly synergistic with mutant but not wild-type p53. Mutant p53 specifically increases TPA induction of VEGF expression without affecting the expression of other TPA-inducible genes. TPA-dependent VEGF expression is also enhanced by human p53 mutated at amino acid 175.

However, Mukhopadhyay et al. [155] found that wild-type p53 downregulated endogenous VEGF mRNA as well as VEGF promoter activity in a dose-dependent manner, whereas mutant p53 had no effect. Overexpression of v-Src, which is known to upregulate VEGF expression, activated expression of a VEGF promoter-luciferase construct in a dose-dependent manner. Moreover, in the presence of wild-type p53, v-Src was unable to activate transcription of the VEGF promoter. Collectively, these data suggest that wild-type p53 plays a role in the suppression of angiogenesis. Adenovirus-mediated wild-type p53 gene transfer also downregulates VEGF expression in human colon cancer cells [185]. The introduction of wild-type p53 into sarcoma cells containing mutant p53 has been shown to significantly reduce VEGF expression. Also, stimulation of endothelial cell migration by conditioned medium from cells expressing mutant p53 is significantly reduced after addition of an anti-VEGF neutralizing antibody to the medium. Using luciferase as the reporter of VEGF promoter activity, we found that wild-type p53 inhibited VEGF promoter activity in SKLMS-1 cells. Deletion analysis defined an 87-bp region (−135 to −48) in the VEGF promoter that is necessary for inhibition of VEGF promoter activity by wild-type p53. Furthermore, the transcription factor Sp1 may be involved in repression of VEGF promoter activity by wild-type p53 in SKLMS-1 cells. These data indicate that wild-type p53 can suppress angiogenesis in human soft-tissue sarcomas via transcriptional repression of VEGF expression [81]. Possible mechanisms include suppression of Sp1 activity by interaction of p53 and promotion of HIF-1α degradation [186–190]. For example, using a human cancer cell line and its derivative from homozygous deletion of the p53 gene via homologous recombination, Ravi et al. [187] found that p53 promotes Mdm2-mediated ubiquitination and proteasomal degradation of HIF-1α. Loss of p53 in tumor cells enhances HIF-1α expression levels and augments HIF-1-dependent transcriptional activation of the VEGF gene in response to hypoxia. Such forced expression of HIF-1α in p53-expressing tumor cells increases hypoxia-induced VEGF expression.

These experimental findings appear to be supported by clinical studies in NSCLC [191], human and noninvasive colorectal carcinoma [192–195], esophageal carcinoma [196], angiosarcoma [197], human breast cancer [198], and gastric carcinoma patients [199]. VEGF is an important angiogenic factor in NSCLC, as its expression is dependent on wild-type p53 loss [200]. Some do not support direct regulation of VEGF by p53 in NSCLC, however [201]. Also, the p53 gene status does not seem to be associated with VEGF expression in oral squamous cell carcinoma [202]. No clinicopathological factors have been significantly correlated with p53 or VEGF expression, and no significant association between p53 and VEGF expression and poor prognosis has been found. In conclusion, p53 and VEGF have not been correlated with prognosis in patients with stage II or III squamous cell carcinoma of the esophagus [203]. In another study, combined analysis of p53 and VEGF showed strong association between the two markers in all 24 liver metastases from colon cancer studied: 9 cases were VEGF and p53 positive, while 15 were VEGF and p53 negative. This relationship was not found in the 34 abdominal metastases studied, which showed concordance between the two markers in nine VEGF- and p53-positive cases only [204]. However, in Hep3B cells stably expressing an inducible p53-estrogen receptor fusion protein and irradiated RKO cells expressing endogenous wild-type p53, VEGF mRNA levels increased in response to hypoxia in both the presence and absence of functional p53. Thus, there was no evidence of a causal relationship between the loss of p53 activity and increased VEGF expression observed during tumor progression. Studies that attribute repressor functions to p53 based on analysis of cells transiently overexpressing this protein should be interpreted cautiously [205].

p73. p73, a newly discovered member of the p53 family, has been cloned and mapped to chromosome 1p36, a region that is frequently deleted in a variety of human cancers. p73 can activate p53-responsive promoters and induce apoptosis when overexpressed in certain p53-deficient tumor cells. Transcriptional silencing of the p73 gene by hypermethylation of a CpG island has been observed in several leukemias and lymphomas, which also show increased expression of VEGF. Additionally, ectopic expression of p73 can downregulate endogenous VEGF gene expression at the mRNA and protein level. This inhibitory effect is mediated by transcriptional repression of the VEGF promoter and involves the promoter region at −85 to −50 bp, which contains a cluster of Sp1 binding sites and two putative Egr-1 binding sites. Mutations in Sp1 binding sites in this region are no longer responsive to p73. In contrast, mutations in the putative Egr-1 binding sites do not influence the p73 responsiveness of the VEGF promoter. This result suggests the involvement of Sp1 binding sites in transcriptional regulation of the VEGF promoter by p73 [83]. However, in a previous study, p73 overexpression correlated with increased production of VEGF mRNA and protein in clonal sublines obtained from a human ovarian carcinoma cell line. The upregulation of VEGF expression was partially due to modulation of the promoter activity and dependent on the p53 status. p73-overexpressing cells are more angiogenic than parental cells, as shown in vitro by their increased chemotactic activity among endothelial cells and in vivo by the generation of more vascularized tumors. These findings indicate a potential role for p73 in tumor angiogenesis [206].

PTEN. PTEN, also referred to as mutated in multiple advanced cancers, was discovered as a tumor suppressor gene and later found to be a phospholipid phosphatase. PTEN negatively regulates Akt activation by preventing its phosphorylation. PTEN therefore inhibits the PI3K/Akt signaling pathway, which is important for cell growth and survival. Overexpression and enhanced activation of PTEN may impair injury healing through at least four mechanisms. Specifically, PTEN can (1) inhibit entry into the cell cycle by inhibiting G1 to S phase progression and arrest cell proliferation required for tissue reconstruction during injury healing, (2) increase apoptosis by blocking Akt activation leading to increased Bad and caspase-9 activities, (3) inhibit hypoxia-induced angiogenesis required for injury healing by blocking Akt-mediated VEGF gene transcription, and (4) inhibit Akt-mediated cell migration, i.e., re-epithelialization, which is also required for injury healing. These same mechanisms can also suppress cancer growth and metastasis. Therefore, elucidating the role of the PTEN/PI3K/Akt pathway will likely advance our knowledge of the mechanisms that control injury healing and cancer growth [207].

Two groups have demonstrated that PTEN regulates hypoxia- and insulin-like growth factor-1 (IGF-1)-induced angiogenic gene expression in glioblastoma-derived cell lines and pathways in human prostate cancer cells by regulating Akt activation of HIF-1α activity. Restoration of wild-type PTEN to glioblastoma cell lines lacking a functional PTEN ablates hypoxia and IGF-1 induction of HIF-1 – regulated genes. In addition, Akt activation leads to HIF-1α stabilization, whereas PTEN attenuates hypoxia-mediated HIF-1α stabilization. Therefore, loss of PTEN deregulates Akt activity and facilitates HIF-1 – mediated VEGF expression [208, 209]. Moreover, an inactivating mutation in the PTEN gene specifically increases the protein level of HIF-1α but not that of HIF-1β in human cancer cell lines. Researchers showed that introduction of wild-type PTEN into the PTEN-negative PC-3 cell line specifically inhibited the expression of HIF-1α but not that of HIF-1β, which is directly correlated with alteration of VEGF reporter gene activity [210].

Inactivation of the PTEN gene and overexpression of the VEGF gene are two common events in some tumors, such as malignant glioma and prostate cancer

[211]. Restoration of PTEN inhibits VEGF expression through downregulation of HIF-1α, a very important transcription factor in the regulation of VEGF expression [208,212,213].

p16. The *Ink4a/Arf*^{4} locus encodes the tumor suppressors $p16^{Ink4a}$ (p16) and $p19^{Arf}$ (Arf) [214]. This locus is often methylated in a broad range of common human solid tumors, including carcinomas of the colon and breast, malignant glioma, such methylation results in functional gene inactivation [215]. Many lines of evidences have demonstrated that p16 is also involved in the regulation of VEGF expression. For example, loss of *p16* associates with the acquisition of angiogenic phenotype in high-grade gliomas, whereas restoration of wild-type p16 into *p16*-deleted glioma cells markedly down-regulates VEGF expression [216]. Moreover, demethylation of $p16^{Ink4a}$ gene results in VEGF downregulation [217]. However, the direct evidence was obtained in *Ink4a/Arf* locus knock out mice with multiple intestinal neoplasia background [218].

Other tumor suppressors. The p63 gene, which was recently identified as a relative of p53, encodes multiple isoforms with structural and functional similarities to and differences from p53. A previous study showed evidence that the two major isoforms of the p63 gene, TAp63γ (p51A) and dNp63α (p73L), repress and upregulate VEGF expression, respectively, at the transcription and protein level. Transient transfection assays have shown that a HIF-1 binding site within the VEGF promoter region is responsible for both upregulation and repression of VEGF promoter activity by dNp63α and TAp63γ, respectively. TAp63γ targets HIF-1α for promoting proteasomal degradation but that dNp63α targets HIF-1α for proteasomal stabilization. Additionally, mammalian two-hybrid assays have shown that HIF-1α-dependent transcription is repressed by TAp63γ as well as p53, whereas it is upregulated by dNp63α in collaboration with the transcription coactivator p300. It has reported that dNp63α acts as a dominant-negative reagent toward both p53- and TAp63γ-mediated degradation of HIF-1α and repression of HIF-1α-dependent transcription. These results suggest that p63 is involved in regulation of VEGF gene expression and that modulation of VEGF expression by TAp63γ and dNp63α is closely correlated with their distinct roles in the regulation of HIF-1α stability [219].

BRCA1. Mutational inactivation of BRCA1 confers increased risk for breast cancer. A recent study showed that BRCA1 and estrogen receptor alpha (ER-α) modulated VEGF gene transcription and secretion in breast cancer cells through interaction with each other [220]. Specifically, ER-α activates VEGF promoter reporter constructs, while BRCA1 inhibits VEGF transactivation. The BRCA1 domain of 1-683 amino acid residues was required for its inhibitory activity. Three mutated forms of BRCA1 (A1708E, M1775R and Y1853X), which have been identified in familial breast cancers, failed to interact with ER-α and to suppress VEGF transactivation. In contrast, overexpression of wild-type BRCA1 in breast cancer cells without functional BRCA1 significantly reduced VEGF expression.

Smad4/DPC4. Smad4/DPC4 is a tumor suppressor gene frequently inactivated in gastrointestinal carcinomas. Restoration of Smad4 into Smad4-null human pancreatic carcinoma cells significantly inhibited VEGF expression [109].

6.5 Signaling Pathways for Inducible VEGF Expression

While the genetic makeup of tumor cells is clearly involved in constitutive VEGF expression by affecting the production and function of transcriptional regulators [128], VEGF expression also can be drastically enhanced by numerous tumor microenvironmental factors, such as hypoxia, acidosis, oxidative stress, and disregulated production of various growth factors and cytokines. Like genetic alterations, a wide variety of tumor microenvironmental stimuli can activate the MAPK and PI3K/Akt pathways [221]. These stimuli include free radicals, hypoxia, acidosis, and a number of growth factors, such as IGF-1, hepatocyte growth factor, fibroblast growth factor (FGF), EGF, VEGF, and PDGF [76, 222–225]. Activation of the MAPK or PI3K/Akt pathway leads to induction of Sp1 transcriptional activity and, at least in part, VEGF upregulation [86, 111, 226].

6.5.1 Hypoxia

Hypoxia, or low oxygen tension, occurs when the growth of neoplastic cells outpaces the rate of new functional blood vessel formation (i.e., inadequate tumor angiogenesis) due to the limitations of oxygen diffusion. As a result, tumor cells lying far from the nearest functional blood vessel experience chronic hypoxia. Because the diffusion distances of glucose and many other critical nutrients are similar to those of oxygen, these cells also experience nutritional deficiencies [227]. Conversely, despite the active angiogenesis often observed in some tumor edges, the overall tumor vasculature is poorly organized and only marginally functional [227–229]. In a normal vascular bed, there is regular, sequential flow of blood through arteries, arterioles, capillaries, postcapillary venules, and veins. However, tumor vascular beds are highly heterogeneous and often disorganized, with arteriolar-venular shunting and other abnormal vessel interconnections. Also, tumor blood vessels are often tortuous in path and irregular in shape and diameter. Moreover, the structure of the vessel walls is often atypical, lacking the smooth muscle elements that are important to regulation of the luminal volume with changes in blood pressure. In addition, invasion and compression by growing tumor cells may result in temporary or permanent collapse of blood vessels and occlusion of individual blood vessels [227–229]. Because of these structural abnormalities in the vasculature, perfusion of tumor tissue is exceedingly chaotic [227–229]. Frequent, transient cessations in microregional blood flow occur both by region and with time, leading to acute hypoxia in tumor cells, even those adjacent to capillaries.

Hypoxia is the most prominent factor for VEGF induction in vitro and in vivo. VEGF was originally identified as the hypoxia-inducible angiogenic factor, because in vitro its mRNA was dramatically induced by exposing cell cultures to hypoxia, and in vivo its expression was higher in tumor cells adjacent to necrotic areas, where they were believed to be under hypoxic conditions [230]. Expression of VEGF mRNA can be induced rapidly and reversibly by hypoxia in many cell types, including normal, transformed, and tumorigenic cells [231, 232]. The mechanisms leading to hypoxic induction of VEGF include transcription-level activation and posttranscription-level stabilization.

Transcription. Hypoxia activates multiple transcription factors. The role of the HIF-1 pathway is best characterized in VEGF induction by hypoxia. Via deletion and mutation analysis, a 28-bp sequence was identified in the promoter region of both rat and human VEGF genes that mediated hypoxia-induced transcription [233, 234]. This sequence showed a high degree of homology with and protein binding characteristics similar to those of the HIF-1 binding site [235]. In the absence of oxygen, HIF-1 binds to HREs, thereby activating VEGF expression. In the presence of oxygen, HIF-1α is bound to the VHL protein. This interaction causes HIF-1α to become ubiquitylated and targeted to the proteasome, where it is degraded [236–238]. Mutations in VHL prevent this ubiquitylation, resulting in an accumulation of HIF-1α and continuous activation of hypoxia-responsive genes [239, 240]. Co-transfection of a HIF-1 expression vector with a reporter containing the VEGF sequences that mediate hypoxia inducibility was shown to significantly upregulate the reporter activity under both hypoxic and normoxic conditions [241]. Also, HIF-1α has been shown to be overexpressed in colon, breast, gastric, lung, skin, ovarian, pancreatic, prostate, and renal carcinoma [242]. Histological analyses have shown that an increased level of intracellular HIF-1α activity is associated with poor prognosis and resistance to therapy in patients with head and neck, ovarian, or esophageal cancer [243].

Posttranscriptional activation. Expression of VEGF mRNA can also be stabilized by hypoxia. Ikeda et al. [244] first described that the VEGF mRNA half-life was increased significantly in C6 rat glioma cells in response to hypoxia treatment. A sequence that mediates this stabilizing effect was mapped to the 3'-UTR of the VEGF gene, and a hypoxia-inducible protein complex was shown to bind to this region in response to hypoxia [113,245]. This process can be blocked by the tyrosine kinase inhibitor genistein, which suggests that the protein that binds to the 3'-UTR may require activation through a tyrosine kinase pathway. On the other hand, another report showed that the 5'-UTR of the VEGF gene also contained destabilizing elements that must co-operate for normoxic instability and hypoxic stability of the mRNA [116].

Translational activation. VEGF expression also appears to be regulated at the translational level under hypoxic conditions. The 5'-UTR of VEGF mRNA, being very G/C rich, forms a complicated structure that is incompatible with efficient translation via ribosomal scanning and the physiological requirement for maximal VEGF production under hypoxic conditions. This region contains a functional internal ribosome entry site that allows translation in an efficient, cap-independent manner where overall translation is reduced and competition for initiation factors is high [120, 246, 247], thereby securing efficient production of VEGF, even under unfavorable stress conditions.

Signal integration. Optimal transcriptional activation of VEGF may require the cooperation of different transcription factors and/or signaling pathways. For example, Damert et al. [110] reported that the binding site of HIF-1 is crucial for hypoxic induction of VEGF gene expression. However, an enhancer subfragment containing this binding site was not sufficient to confer full responsiveness to hypoxia. Addition of upstream sequences restored the full sensitivity to hypoxic induction. This potentiating effect is due to AP-1 binding. The "potentiating" sequences (AP-1 transcription factor binding sites) are unable to confer hypoxia responsiveness on their own, but they do potentiate hypoxia induction of the human VEGF gene via HIF-1. AP-1 is a transcription factor belonging to the leucine zipper family. This factor is a dimer composed of jun/jun (c-jun, junB, and junD) or jun/fos subunits (s-fos, fosB, fra-1, and fra-2). It has been demonstrated that environmental stresses such as acidosis, ultraviolet irradiation, and cytokines can induce the synthesis of jun and fos proteins, leading to increased AP-1 binding activity. Furthermore, phosphorylation of jun by JNK and/or fos by the FRK MAPK is required for AP-1 transcriptional activity. Thus, many tumor microenvironmental factors may affect VEGF expression through the signaling pathways that can lead to activation of AP-1 [104, 105]. A similar example was reported by Brenneisen [112] showing that full induction of VEGF expression stimulated by ultraviolet B radiation requires the cooperation of Sp1 with AP-1. While transforming growth factor-β (TGF-β) and hypoxia pathways can synergize in the regulation of VEGF expression at the transcriptional level, this cooperation has been mapped on the human VEGF promoter within a region at -1006 to -954 bp that contains functional DNA binding sequences for HIF-1 and Smads, which was confirmed by demonstration of physical interaction between Samd and HIF-1α [71].

6.5.2 *Acidosis*

With the development of hypoxic regions within a solid tumor, the anaerobic metabolism of tumor cells and production of acidic metabolites increases. Furthermore, reduced blood flow hinders the removal of these metabolites. Consequently, hydrogen ions accumulate and cause a decrease in extracellular pH levels [248, 249]. Numerous bodies of experimental evidence have supported this conclusion [227, 248, 249]. The remarkable temporal and regional variation in extracellular pH levels may reflect temporal and regional heterogeneity of blood perfusion and tumor cell metabolism [228, 248, 249]. Acidic tumor pH has been implicated in the regulation of many aspects of tumor biology. For example, tumor acidosis regulates VEGF expression and tumor angiogenesis. Fukumura et al. [250] developed a novel in vivo microscopy technique to simultaneously measure VEGF promoter activity, pO_2, and pH and found that, under hypoxic or neutral pH conditions, VEGF promoter activity in human brain tumors increased, while pO_2 decreased, both independent of pH. Under low pH and oxygenated conditions, VEGF promoter activity increased, while the pH level decreased, both independent of pO_2. In agreement with the in vivo findings, both hypoxia and acidic pH induced VEGF expression in these cells in vitro and showed no additive effect with combined hypoxia and low pH. These results suggest that VEGF upregulation in brain tumors is regulated by both tissue pO_2 and pH via distinct pathways. The same group further investigated the mechanism of how low pH induces VEGF expression. They performed in vitro experiments by replacing the neutral culture medium (pH 7.3) with acidic pH medium (pH 6.6). They found that acidic pH upregulated VEGF mRNA and protein production in human glioblastoma cells. Functional analysis of the VEGF promoter revealed that the sequence from −961 to −683 bp upstream of the transcription start site was responsible for the transcriptional activation of the VEGF gene by acidic pH. This region contains the binding site for AP-1. They also revealed that acidic pH activates Ras and the ERK1/2 MAPK pathway and that increased AP-1 transcriptional activity is responsible for the upregulation of VEGF expression by low pH [251]. However, Brooks et al. [252] found that pH (range, 7.0–8.0) did not significantly affect VEGF production under normoxic conditions, whereas decreased pH inhibited VEGF expression in rat C6 glioma cells, retinal Müller cells [252], and osteoblasts [253]. D'Arcangelo et al. [254] recently indicated that although acidosis induced VEGF as well as basic FGF (bFGF) expression in bovine endothelial cells, increased secretion of VEGF and bFGF failed to induce endothelial cell migration or morphogenesis.

We have determined the role of low tumor pH in the expression and regulation of VEGF in various types of human tumor cells using various techniques. Specifically, the expression of VEGF mRNA increased when tumor cells were treated for 6 h in culture medium at pH 7.1, 6.9, or 6.7 as compared with that at pH 7.3. Longer incubation times led to further elevation of VEGF mRNA expression at pH 7.1 and 6.9 but significantly decreased expression at pH 6.7 and lower. Consistent with the elevation of VEGF mRNA expression, the acidic culture medium led to increased VEGF secretion. Therefore, transient exposure to acidosis at a pH near the neutral level upregulates VEGF expression; however, extensive exposure to mild or intensive acidosis inhibited VEGF expression. These findings may explain why different pH levels lead to diverse expression of VEGF [252, 254–256].

Apparently, pH regulated VEGF expression at both the transcriptional and post-transcriptional level in the studies described above. At a near-neutral pH level, VEGF expression was elevated due to both activation of transcription and increased transcript stability, whereas at a mildly acidic pH level, increased VEGF expression was mostly due to increased VEGF transcript stability and, to a much lesser extent, increased gene transcription. Interestingly, a low pH level that did not activate but rather inhibited transcription did enhance transcript stability. The increased VEGF mRNA stability was consistent with reports showing that there are stability-sensitive elements in the 3'-UTR of the VEGF transcript [257–259]. However, it remains to be determined whether acidosis acts upon these elements through the same factors. On the other hand, we identified two NF-κB binding sites on the VEGF 3'-UTR [260]. Transient exposure to pH 7.1 and 6.9 increased NF-κB binding to the NF-κB site of the VEGF gene, whereas prolonged exposure did the opposite. This finding is consistent with several reports showing that dominant-negative IκBα downregulates VEGF expression [261, 262]. Because acidosis also activates AP-1 binding to AP-1 sites of the VEGF gene, it is highly possible that the activation and cooperation of the NF-κB and AP-1 binding sites may contribute to the activation of VEGF transcription by a low extracellular pH level.

6.5.3 Reactive Nitrogen and Oxygen Species

Free radicals are an integral part of metabolism, and continuous production of them is ubiquitous in all respiring organisms. Normally, free radicals are neutralized by enzymatic activity or natural antioxidants. Thus, generation of free radicals does not pose danger so long as the rates of free radical production and eradication remain in balance [263]. Since the first description of free radicals in the 1950s, substantial achievements in free radical research have been made that are the bases of the current explosion of interest in the new fields of cancer etiology and chemoprevention. Stress from many sources leads to increased production of free radicals and associated reactive oxygen and nitrogen species. In addition, many disease states may be aggravated by the presence of free radicals. It is known that free radical production is enhanced in tumors, which may be directly related to tumor oxygen supply and metabolism [263, 264]. As signaling molecules, many free radicals regulate the expression of genes important to tumor angiogenesis, including VEGF.

Reactive nitrogen species. NO was discovered to be a potent vasodilator in 1979, and later identified as an endothelium-derived relaxing factor. NO is synthesized from L-arginine by three isoenzymes called NO synthases (NOSs), including neuronal NOS (NOS I), inducible NOS (NOS II), and endothelial NOS (NOS III). Generally, NOS I and III are constitutively expressed and calcium dependent, while NOS II is inducible and calcium independent. It has been recognized that NO plays significant roles in tumor growth and metastasis [265]. One of its major functions is regulation of tumor angiogenesis. Because increased NOS II and VEGF expression levels have frequently been detected in human breast, brain, head and neck, colon, and gastric cancers and hepatocellular carcinoma, it appears that there is a direct relationship between NO production and VEGF expression [266–268]. However, this relationship is more complicated than just upregulation of each other [269] or reciprocal regulation among one another [270]. Concomitant NOS II and VEGF overexpression may be simply due to a hypoxic and/or acidic tumor microenvironment [266, 267]. In a hypoxic tumor environment, HIF-1 and thus NOS II and VEGF are highly expressed, while the negative feedback of NO production on VEGF expression may be overwhelmed by the strong induction of HIF-1 activity.

Some evidence indicates that NO upregulates VEGF. For example, in an in vivo animal model, administration of an NOS inhibitor, N(omega)-nitro-L-argine methyl ester (L-NAME), significantly suppressed tumor growth as well as markedly decreased VEGF levels and reduced neovascularization in tumor tissues [271]. Upregulation of VEGF by NO was also observed in keratinocytes during wound repair using the NOS inhibitors N^G-monomethyl-L-arginine and L-N(6)-(1-iminoethyl)lysine [272, 273], in rat mesangial cells using S-nitroso-glutathione (GSNO) as a NO donor [274], and in rat smooth muscle cells using L-NAME and the GTP cyclohydrolase I inhibitor diaminohydroxypyrimidine [275]. More convincing evidence supporting the upregulation of VEGF expression by NO comes from studies of gene knockout mice. VEGF expression induced by cytokines was significantly attenuated in macrophages from NOS $II^{-/-}$ mice [276], tumors growing in NOS $II^{-/-}$ mice displayed lower VEGF expression [277].

Treatment of human glioblastoma and hepatocellular carcinoma cells with the NO donor S-nitroso-N-acetyl-D, L-penicillamine increases VEGF mRNA expression and prolongs the mRNA half-life, suggesting that NO regulates VEGF expression at both the transcriptional and posttranscriptional level [278]. There are several findings that support this notion. For example, NO donors induce HIF-1α protein accumulation and may enhance VEGF expression through HRE in the promoter of the VEGF gene [279,280] in a manner similar to that of hypoxia [72,281]. Because NO activates the p42/p44 MAPK signaling pathway [282], NO may also upregulate VEGF expression through increasing Sp1 transcriptional activity [283]. Moreover, NO-mediated VEGF upregulation is potentiated by the AP-1 element, which is located next to the HRE in the VEGF promoter [72], although another report has shown that NO can inhibit binding of AP-1 to the VEGF promoter [284].

Other evidence indicates that NO downregulates VEGF expression. The expression of VEGF and VEGF receptor transcripts is clearly decreased by the NO donor sodium nitroprusside (SNP) and increased by L-NAME in isolated perfused rat lung [285]. Another group found similar downregulation of VEGF expression by NO in bovine pulmonary artery endothelial cells and rat aorta smooth muscle cells [286]. Using GSNO and cGMP analogues and NO downstream signal molecules, the authors showed that NO repressed the hypoxic induction of VEGF expression at the transcriptional level by decreasing HIF-1 DNA binding activity without decreasing the HIF-1 protein levels [286]. Also, Yin et al. [287] demonstrated that NOS II expression inhibits HIF-1 activity under hypoxic conditions in C6 glioma cells transfected with an NOS II gene and a VEGF-promoter-driven luciferase gene. HIF-1 induction of VEGF-luciferase activity in C6 cells is also inhibited by SNP. Furthermore, pretreatment of C6 cells with N-acetyl-l-cysteine, an antioxidant, nullified the inhibitory effect of NOS II on HIF-1 binding. NO inhibition of hypoxia-induced VEGF expression was also observed in immortalized human retina epithelial cells, H-Ras-transfected murine capillary endothelial cells, and NF-κB knockout 3T3 fibroblasts using SNP as an NO donor [288]. TPA- and Ras-transfection-induced VEGF expression were also inhibited by NO, suggesting that NO serves as an endogenous inhibitor of both hypoxia- and non-hypoxia-enhanced VEGF expression in vivo.

The apparent discrepancy may have many causes, including the use of different NOS inducers/inhibitors and NO donors, use of different cell lines, and intensity of NO exposure [289]. For example, NO donors such as S-nitroso-N-acetyl-D, L-penicillamine, GSNO, and SNP have very different chemistries and a variety of biological effects besides donation of NO. Other causes, such as the use of different cell systems, may also account for this discrepancy. Indeed, most of the inhibitory effects of NO on VEGF expression have been observed in endothelial and smooth muscle cells, while the upregulation effect has been observed mostly in tumor cells.

One may speculate that normal cells have an intact regulatory pathway for NO and VEGF, which inhibits regulation, while malignant and transformed cells somehow reverse or disconnect the regulation pathway for the benefit of their growth. It has been observed that NOS II-transfected p53 mutant colon cancer cell lines have higher expression of VEGF and higher tumorigenic potential than do cells with wild-type p53 [290]. Another explanation for the discrepant results is the level of NO production. A high NO concentration may downregulate VEGF expression, whereas a low NO concentration may upregulate it. However, in an isolated aorta model, a low NO concentration was sufficient to inhibit VEGF expression [291]. Further studies are clearly needed to elucidate the definitive role of NO in the regulation of VEGF as well as angiogenesis and tumor growth.

Reactive oxygen species (ROS). ROS, which result in an oxidative challenge to cells, can be exogenous or endogenous. The major sources of cell ROS, such as superoxide O_2, hydroxyl radical OH, and H_2O_2, are continuously generated as products of cellular mitochondrial metabolism [292]. ROS can also be generated upon receptor activation either by nicotinamide adenine dinucleotide phosphate oxidase-like membrane complexes [293] or through the oxidative metabolism of free arachidonic acid released by ligand-dependent phospholipases [294], which have a major role in transducing intracellular signals by activated growth factor receptors. Cellular receptors for EGF, PDGF, insulin, and immunological stimuli are in fact all linked to ROS-generating systems, the blockage of which, as well as the removal of oxygen radicals by chemical and enzyme antioxidants, severely compromises cell response to mitogenic stimulation [295–297]. Therefore, ROS may serve as an intracellular messenger following receptor activation by a variety of bioactive peptides, including growth factors, cytokines, and hormones [298, 299].

An increased level of ROS has been reported to participate in malignant transformation as well as metastasis [292, 300, 301]. Two of the main mechanisms of ROS involved in tumor development and progression are induction of VEGF expression and promotion of angiogenesis [300, 302–304]. It appears that cellular response to ROS is mediated mainly by activation of the Ras/raf/p42/p44 MAPK, Rac/MEKK1/JNKs, and Rac/PAC/p38 MAPK signaling pathways, resulting in activation of many downstream targets, such as Sp1, AP-1, NF-κB, and Stat3 [86, 305–308], which are important for regulation of VEGF expression. In addition, ROS activates the PI3K/Akt pathway, which leads to HIF-1 induction and VEGF upregulation [309–311]. Interestingly, a recent study demonstrated that ROS also functions as a downstream mediator of angiogenic signaling by VEGF/VEGGR2 [312].

6.5.4 Growth Factors and Cytokines

VEGF expression can be regulated by a number of growth factors and cytokines. For example, EGF stimulates VEGF release by glioblastoma cells. Also, EGF, TGF-β [69], and keratinocyte growth factor significantly induce VEGF mRNA expression in keratinocytes [313]. It has been proposed that VEGF may be a paracrine mediator for indirectly acting angiogenic agents, such as TGF-β [314]. Other growth factors, such as bFGF [315], FGF4 [316], PDGF [317], and hepatocyte growth factor [318], can also induce VEGF expression. Furthermore, cytokines can induce VEGF expression in different cell types. It has been shown that interleukin-1β induces VEGF expression in rat aortic smooth muscle cells [319] and that interleukin-1α induces VEGF expression in cultured synovial fibroblasts [320]. Furthermore, interleukin-6 induces VEGF expression in a variety of cells [321], and TNF-α [77] induces VEGF expression in glioma cells. IGF-1, a mitogen implicated in the growth of several malignancies, has also been shown to induce VEGF mRNA and protein expression in colorectal carcinoma cells [322]. The mechanism of this induction involves both gene transcription activation and mRNA stabilization.

However, interferons (IFNs) play different roles in the regulation of VEGF expression. IFN-γ, which is one of the major inflammatory cytokines besides interleukin-1 and TNF, induces VEGF expression in Kaposi sarcoma [323], hepatocelluar carcinoma [324], and melanoma cells [325] and keratinocytes [273]. Macrophages activated by IFN-γ and lipopolysaccharide have also shown elevated expression of VEGF [326]. On the other hand, IFN-β decreases VEGF expression in gastric carcinoma [327] and melanoma cells [325], and IFN-α suppresses VEGF production in human peripheral blood mononuclear cells [328] and human neuroendocrine tumors [329]. Clinical trials using IFN-α in patients with renal cell carcinoma have demonstrated that it inhibits the serum level of VEGF. Therefore, IFN-α may confer its antitumor activity, at least in part, through its antiangiogenic activity, which results from Sp1- and/or Sp3-mediated inhibition of VEGF gene transcription [329, 330].

6.5.5 Others

Many other stimuli can regulate VEGF expression. These include ultraviolet B radiation [107, 112], ionizing radiation [331], serum starvation [332], prostaglandins [333–335], okadaic acid [336], glucose deficiency [257, 337, 338], and iron chelators [339]. Many of these stimuli can be deemed cellular stress signals. Arguably, VEGF expression induced by these stress signals may not function as an angiogenic factor but rather a surviving factor for endothelial cells and preexisting vessels [340–343].

In summary, both genetic and epigenetic factors may use one or more pathways to affect VEGF expression at different levels, such as transcriptional activation, mRNA stabilization, and translational regulation. Different tumor microenvironmental stimuli may activate different signaling pathways, thus inducing different regulatory factors to modulate VEGF expression. Several transcription factors have been demonstrated to be critical to the transcriptional regulation of VEGF expression. The transcriptional activities of these transcription factors are mainly subject to the Ras-raf-extracellular signal-regulated kinase-p42/p44 MAPK and PI3K/Akt signaling pathways. Some genetic and epigenetic factors may activate both of these as well as other signal transduction pathways, while the full induction of VEGF expression may require the activation and cooperation of several transcription factors.

6.6 Concluding Remarks

Angiogenic switch-on is crucial for sustained growth and metastasis of most tumors. There is now little doubt that VEGF is a crucial factor in the regulation of tumor angiogenesis and is a very important target for antitumor therapy. However, VEGF-mediated angiogenesis involves a plethora of modifiers in the tumor microenvironment and intricate mechanisms of expression. The extent and pattern of tumor angiogenesis are profoundly influenced by the accumulation of genetic abnormalities in tumor cells and microenvironmental factors, crosstalk among the various molecular and cellular components of the tumor environment, and establishment of VEGF gradients between tumor cells and the surrounding donor vasculature. Their relative contribution is likely to change with the tumor type, site, and progression path. Presumably, VEGF-mediated tumor angiogenic switch-on is regulated by the synergistic cooperation of two signals: the genetic makeup of tumor cells that signals constitutive VEGF expression (constitutive angiogenic signal) and of tumor microenvironmental stimuli that signal inducible VEGF expression (inducible angiogenic signal). The constitutive VEGF signal, which stems from various genetic alterations such as loss of function of tumor suppressor genes and/or gain of function of oncogenes, is a prerequisite and acts as an initiation signal to create a preangiogenic condition for tumor angiogenesis. With an increase in tumor mass and tumor-host interaction comes inducible VEGF expression from both tumor and tumor stromal cells by various stimuli, especially hypoxia, acidosis, and free radicals. This inducible signal cooperates with the constitutive signal, leading to a high level of VEGF production and then tumor angiogenic switch-on. It is likely that the constitutive VEGF signal creates the preangiogenic conditions (initiation status) that is essential for tumor angiogenesis, but that tumor angiogenic switch-on (robust status) is dependent on the synergistic cooperation between the constitutive and inducible VEGF signal. This hypothesis incorporates quantitative data on the magnitude and temporal sequence of VEGF expression and establishes a framework for VEGF-mediated tumor angiogenic switch-on. However, numerous critical questions remain unanswered. For example, what is the qualitative and quantitative difference in VEGF-mediated tumor angiogenesis in experimental versus clinical tumors? Also, how does VEGF cooperate with other growth factors and cytokines to form an angiogenic cocktail in different angiogenic situations? Furthermore, how does the interplay between genetic and microenvironmental factors influence VEGF-mediated tumor angiogenesis? A better understanding of the signaling pathways and their interaction and/or integration will facilitate the development of more effective therapeutic strategies by targeting VEGF-mediated tumor angiogenesis.

Acknowledgments This work was supported in part by Research Project Grant #RPG-00-054-01-CMS and Research Scholar Grant #CSM-106640 and #CSM-106659 from the American Cancer Society; grant 1R01-CA093829 and Cancer Center Support Core Grant CA 16672 from the National Cancer Institute, National Institutes of Health; and grant LF-4 from the Lustgarten Foundation for Pancreatic Cancer Research (K. X.). We thank Judy King for expert help in the preparation of this manuscript and Don Norwood for editorial comments.

References

1. Senger DR, Galli SJ, Dvorak AM, Perruzzi CA, Harvey VS, Dvorak HF. Tumor cells secrete a vascular permeability factor that promotes accumulation of ascites fluid. Science. 1983; 219(4587): 983–85.
2. Leung DW, Cachianes G, Kuang WJ, Goeddel DV, Ferrara N. Vascular endothelial growth factor is a secreted angiogenic mitogen. Science. 1989; 246: 1306–09.
3. Keck PJ, Hauser SD, Krivi G, Sanzo K, Warren T, Feder J, Connolly DT. Vascular permeability factor, an endothelial cell mitogen related to PDGF. Science. 1989; 246: 1309–12.
4. Baldwin ME, Stacker SA, Achen MG. Molecular control of lymphangiogenesis. Bioessays. 2002; 24: 1030–40.
5. Nagy JA, Vasile E, Feng D, Sundberg C, Brown LF, Detmar MJ, Lawitts JA, Benjamin L, Tan X, Manseau EJ, Dvorak AM, Dvorak HF. Vascular permeability factor/vascular endothelial growth factor induces lymphangiogenesis as well as angiogenesis. J Exp Med. 2002; 196: 1497–506.
6. Ohm JE, Gabrilovich DI, Sempowski GD, Kisseleva E, Parman KS, Nadaf S, Carbone DP. VEGF inhibits T-cell development and may contribute to tumor-induced immune suppression. Blood. 2003; 101: 4878–86.
7. Ohm JE, Carbone DP. VEGF as a mediator of tumor-associated immunodeficiency. Immunol Res. 2001; 23: 263–72.
8. Lyden D, Hattori K, Dias S, Costa C, Blaikie P, Butros L, Chadburn A, Heissig B, Marks W, Witte L, Wu Y, Hicklin D, Zhu Z, Hackett NR, Crystal RG, Moore MA, Hajjar KA, Manova K, Benezra R, Rafii S. Impaired recruitment of bone-marrow-derived endothelial and hematopoietic precursor cells blocks tumor angiogenesis and growth. Nat Med. 2001; 7: 1194–201.
9. Rafii S, Lyden D, Benezra R, Hattori K, Heissig B. Vascular and haematopoietic stem cells: novel targets for anti-angiogenesis therapy? Nat Rev Cancer. 2002; 2: 826–35.
10. Mann DR, Plant TM. Leptin and pubertal development. Semin Reprod Med. 2002; 20: 93–102.
11. Mattei MG, Borg JP, Rosnet O, Marme D, Birnbaum D. Assignment of vascular endothelial growth factor (VEGF) and placenta growth factor (PLGF) genes to human chromosome 6p12-p21 and 14q24-q31 regions, respectively. Genomics. 1996; 32: 168–69.
12. Paavonen K, Horelli-Kuitunen N, Chilov D, Kukk E, Pennanen S, Kallioniemi OP, Pajusola K, Olofsson B, Eriksson U, Joukov V, Palotie A, Alitalo K. Novel human vascular endothelial growth factor genes VEGF-B and VEGF-C localize to chromosomes 11q13 and 4q34, respectively. Circulation. 1996; 93: 1079–82.
13. Joukov V, Pajusola K, Kaipainen A, Chilov D, Lahtinen I, Kukk E, Saksela O, Kalkkinen N, Alitalo K. A novel vascular endothelial growth factor, VEGF-C, is a ligand for the Flt4 (VEGFR-3) and KDR (VEGFR-2) receptor tyrosine kinases. EMBO J. 1996 Jan 15; 15(2): 290–98. Erratum in: EMBO J. 1996; 15: 1751.
14. Yamada Y, Nezu J, Shimane M, Hirata Y. Molecular cloning of a novel vascular endothelial growth factor, VEGF-D. Genomics. 1997; 42: 483–88.
15. Lyttle DJ, Fraser KM, Fleming SB, Mercer AA, Robinson AJ. Homologs of vascular endothelial growth factor are encoded by the poxvirus orf virus. J Virol. 1994; 68: 84–92.

16. Maglione D, Guerriero V, Viglietto G, Delli-Bovi P, Persico MG. Isolation of a human placenta cDNA coding for a protein related to the vascular permeability factor. Proc Natl Acad Sci USA. 1991; 88: 9267–71.
17. Olofsson B, Korpelainen E, Pepper MS, Mandriota SJ, Aase K, Kumar V, Gunji Y, Jeltsch MM, Shibuya M, Alitalo K, Eriksson U. Vascular endothelial growth factor B (VEGF-B) binds to VEGF receptor-1 and regulates plasminogen activator activity in endothelial cells. Proc Natl Acad Sci USA. 1998; 95: 11709–14.
18. Makinen T, Olofsson B, Karpanen T, Hellman U, Soker S, Klagsbrun M, Eriksson U, Alitalo K. Differential binding of vascular endothelial growth factor B splice and proteolytic isoforms to neuropilin-1. J Biol Chem. 1999; 274: 21217–22.
19. Achen MG, Jeltsch M, Kukk E, Makinen T, Vitali A, Wilks AF, Alitalo K, Stacker SA. Vascular endothelial growth factor D (VEGF-D) is a ligand for the tyrosine kinases VEGF receptor 2 (Flk1) and VEGF receptor 3 (Flt4). Proc Natl Acad Sci USA. 1998; 95: 548–53.
20. Stacker SA, Stenvers K, Caesar C, Vitali A, Domagala T, Nice E, Roufail S, Simpson RJ, Moritz R, Karpanen T, Alitalo K, Achen MG. Biosynthesis of vascular endothelial growth factor-D involves proteolytic processing which generates non-covalent homodimers. J Biol Chem. 1999; 274: 32127–36.
21. Wise LM, Veikkola T, Mercer AA, Savory LJ, Fleming SB, Caesar C, Vitali A, Makinen T, Alitalo K, Stacker SA. Vascular endothelial growth factor (VEGF)-like protein from orf virus NZ2 binds to VEGFR2 and neuropilin-1. Proc Natl Acad Sci USA. 1999; 96: 3071–76.
22. Cao Y, Ji WR, Qi P, Rosin A, Cao Y. Placenta growth factor: identification and characterization of a novel isoform generated by RNA alternative splicing. Biochem Biophys Res Commun. 1997; 235: 493–98.
23. Park JE, Chen HH, Winer J, Houck KA, Ferrara N. Placenta growth factor: potentiation of vascular endothelial growth factor bioactivity, in vitro and in vivo, and high affinity binding to Flt-1 but not to Flk-1/KDR. J Biol Chem. 1994; 269: 25646–54.
24. DiSalvo J, Bayne ML, Conn G, Kwok PW, Trivedi PG, Soderman DD, Palisi TM, Sullivan KA, Thomas KA. Purification and characterization of a naturally occurring vascular endothelial growth factor: placenta growth factor heterodimer. J Biol Chem. 1995; 270: 7717–23.
25. Claesson-Welsh L. Signal transduction by vascular endothelial growth factor receptors. Biochem Soc Trans. 2003; 31(Pt 1): 20–24.
26. Soker S, Takashima S, Miao HQ, Neufeld G, Klagsbrun M. Neuropilin-1 is expressed by endothelial and tumor cells as an isoform-specific receptor for vascular endothelial growth factor. Cell. 1998; 92: 735–45.
27. Stacker SA, Baldwin ME, Achen MG. The role of tumor lymphangiogenesis in metastatic spread. FASEB J. 2002; 16: 922–34.
28. Zachary I, Gliki G. Signaling transduction mechanisms mediating biological actions of the vascular endothelial growth factor family. Cardiovasc Res. 2001; 49: 568–81.
29. Karkkainen MJ, Saaristo A, Jussila L, Karila KA, Lawrence EC, Pajusola K, Bueler H, Eichmann A, Kauppinen R, Kettunen MI, Yla-Herttuala S, Finegold DN, Ferrell RE, Alitalo K. A model for gene therapy of human hereditary lymphedema. Proc Natl Acad Sci USA. 2001; 98: 12677–82.
30. Ferrara N, Henzel WJ. Pituitary follicular cells secrete a novel heparin-binding growth factor specific for vascular endothelial cells. Biochem Biophys Res Commun. 1989; 161: 851–58.
31. Olofsson B, Pajusola K, Kaipainen A, von Euler G, Joukov V, Saksela O, Orpana A, Pettersson RF, Alitalo K, Eriksson U. Vascular endothelial growth factor B, a novel growth factor for endothelial cells. Proc Natl Acad Sci USA. 1996; 93: 2576–81.
32. Matsumoto T, Claesson-Welsh L. VEGF receptor signal transduction. Sci STKE. 2001; 2001(112): RE21.
33. Zachary I. Signaling mechanisms mediating vascular protective actions of vascular endothelial growth factor. Am J Physiol Cell Physiol. 2001; 280: C1375–86.
34. Cross MJ, Claesson-Welsh L. FGF and VEGF function in angiogenesis: signalling pathways, biological responses and therapeutic inhibition. Trends Pharmacol Sci. 2001; 22: 201–7.

35. Almand B, Resser JR, Lindman B, Nadaf S, Clark JI, Kwon ED, Carbone DP, Gabrilovich DI. Clinical significance of defective dendritic cell differentiation in cancer. Clin Cancer Res. 2000; 6(5): 1755–66.
36. Saito H, Tsujitani S, Ikeguchi M, Maeta M, Kaibara N. Relationship between the expression of vascular endothelial growth factor and the density of dendritic cells in gastric adenocarcinoma tissue. Br J Cancer. 1998; 78: 1573–7.
37. Gabrilovich D, Ishida T, Oyama T, Ran S, Kravtsov V, Nadaf S, Carbone DP. Vascular endothelial growth factor inhibits the development of dendritic cells and dramatically affects the differentiation of multiple hematopoietic lineages in vivo. Blood. 1998; 92: 4150–66.
38. Oyama T, Ran S, Ishida T, Nadaf S, Kerr L, Carbone DP, Gabrilovich DI. Vascular endothelial growth factor affects dendritic cell maturation through the inhibition of nuclear factor-kappa B activation in hemopoietic progenitor cells. J Immunol. 1998; 160: 1224–32.
39. Gabrilovich DI, Chen HL, Girgis KR, Cunningham HT, Meny GM, Nadaf S, Kavanaugh D, Carbone DP. Production of vascular endothelial growth factor by human tumors inhibits the functional maturation of dendritic cells. Nat Med. 1996; 2: 1096–103.
40. Gabrilovich DI, Ishida T, Nadaf S, Ohm JE, Carbone DP. Antibodies to vascular endothelial growth factor enhance the efficacy of cancer immunotherapy by improving endogenous dendritic cell function. Clin Cancer Res. 1999; 5: 2963–70.
41. Bouloumie A, Schini-Kerth VB, Busse R. Vascular endothelial growth factor up-regulates nitric oxide synthase expression in endothelial cells. Cardiovasc Res. 1999; 41: 773–80.
42. Shen BQ, Lee DY, Zioncheck TF. Vascular endothelial growth factor governs endothelial nitric-oxide synthase expression via a KDR/Flk-1 receptor and a protein kinase C signaling pathway. J Biol Chem. 1999; 274: 33057–63.
43. Pepper MS, Ferrara N, Orci L, Montesano R. Vascular endothelial growth factor (VEGF) induces plasminogen activators and plasminogen activator inhibitor-1 in microvascular endothelial cells. Biochem Biophys Res Commun. 1991; 181: 902–06.
44. Landgren E, Schiller P, Cao Y, Claesson-Welsh L. Placenta growth factor stimulates MAP kinase and mitogenicity but not phospholipase C-gamma and migration of endothelial cells expressing Flt 1. Oncogene. 1998; 16: 359–67.
45. Mason JC, Lidington EA, Yarwood H, Lublin DM, Haskard DO. Induction of endothelial cell decay-accelerating factor by vascular endothelial growth factor: a mechanism for cytoprotection against complement-mediated injury during inflammatory angiogenesis. Arthritis Rheum. 2001; 44: 138–50.
46. Zebrowski BK, Yano S, Liu W, Shaheen RM, Hicklin DJ, Putnam JB Jr, Ellis LM. Vascular endothelial growth factor levels and induction of permeability in malignant pleural effusions. Clin Cancer Res. 1999; 5: 3364–68.
47. Zebrowski BK, Liu W, Ramirez K, Akagi Y, Mills GB, Ellis LM. Markedly elevated levels of vascular endothelial growth factor in malignant ascites. Ann Surg Oncol. 1999; 6: 373–78.
48. Hazelton D, Nicosia RF, Nicosia SV. Vascular endothelial growth factor levels in ovarian cyst fluid correlate with malignancy. Clin Cancer Res. 1999; 5: 823–29.
49. Minagawa N, Nakayama Y, Hirata K, Onitsuka K, Inoue Y, Nagata N, Itoh H. Correlation of plasma level and immunohistochemical expression of vascular endothelial growth factor in patients with advanced colorectal cancer. Anticancer Res. 2002; 22: 2957–63.
50. Chao C, Al-Saleem T, Brooks JJ, Rogatko A, Kraybill WG, Eisenberg B. Vascular endothelial growth factor and soft tissue sarcomas: tumor expression correlates with grade. Ann Surg Oncol. 2001; 8: 260–67.
51. Lamszus K, Lengler U, Schmidt NO, Stavrou D, Ergun S, Westphal M. Vascular endothelial growth factor, hepatocyte growth factor/scatter factor, basic fibroblast growth factor, and placenta growth factor in human meningiomas and their relation to angiogenesis and malignancy. Neurosurgery. 2000; 46: 938–47.
52. Dvorak HF. Vascular permeability factor/vascular endothelial growth factor: a critical cytokine in tumor angiogenesis and a potential target for diagnosis and therapy. J Clin Oncol. 2002; 20: 4368–80.

53. Toi M, Matsumoto T, Bando H. Vascular endothelial growth factor: its prognostic, predictive, and therapeutic implications. Lancet Oncol. 2001; 2: 667–73.
54. Gasparini G, Toi M, Gion M, Verderio P, Dittadi R, Hanatani M, Matsubara I, Vinante O, Bonoldi E, Boracchi P, Gatti C, Suzuki H, Tominaga T. Prognostic significance of vascular endothelial growth factor protein in node-negative breast carcinoma. J Natl Cancer Inst. 1997 Jan 15; 89(2): 139–47.
55. Yamamoto S, Konishi I, Mandai M, Kuroda H, Komatsu T, Nanbu K, Sakahara H, Mori T. Expression of vascular endothelial growth factor (VEGF) in epithelial ovarian neoplasms: correlation with clinicopathology and patient survival, and analysis of serum VEGF levels. Br J Cancer. 1997; 76: 1221–27.
56. Abdulrauf SI, Edvardsen K, Ho KL, Yang XY, Rock JP, Rosenblum ML. Vascular endothelial growth factor expression and vascular density as prognostic markers of survival in patients with low-grade astrocytoma. J Neurosurg. 1998; 88: 513–20.
57. Ishigami SI, Arii S, Furutani M, Niwano M, Harada T, Mizumoto M, Mori A, Onodera H, Imamura M. Predictive value of vascular endothelial growth factor (VEGF) in metastasis and prognosis of human colorectal cancer. Br J Cancer. 1998; 78: 1379–84.
58. Yuan A, Yu CJ, Chen WJ, Lin FY, Kuo SH, Luh KT, Yang PC. Correlation of total VEGF mRNA and protein expression with histologic type, tumor angiogenesis, patient survival and timing of relapse in non-small-cell lung cancer. Int J Cancer. 2000; 89: 475–83.
59. Seo Y, Baba H, Fukuda T, Takashima M, Sugimachi K. High expression of vascular endothelial growth factor is associated with liver metastasis and a poor prognosis for patients with ductal pancreatic adenocarcinoma. Cancer. 2000; 88: 2239–45.
60. Tempfer C, Obermair A, Hefler L, Haeusler G, Gitsch G, Kainz C. Vascular endothelial growth factor serum concentrations in ovarian cancer. Obstet Gynecol. 1998; 92: 360–63.
61. Hyodo I, Doi T, Endo H, Hosokawa Y, Nishikawa Y, Tanimizu M, Jinno K, Kotani Y. Clinical significance of plasma vascular endothelial growth factor in gastrointestinal cancer. Eur J Cancer. 1998; 34: 2041–45.
62. Karayiannakis AJ, Syrigos KN, Polychronidis A, Zbar A, Kouraklis G, Simopoulos C, Karatzas G. Circulating VEGF levels in the serum of gastric cancer patients: correlation with pathological variables, patient survival, and tumor surgery. Ann Surg. 2002; 236: 37–42.
63. Aguayo A, Kantarjian HM, Estey EH, Giles FJ, Verstovsek S, Manshouri T, Gidel C, O'Brien S, Keating MJ, Albitar M. Plasma vascular endothelial growth factor levels have prognostic significance in patients with acute myeloid leukemia but not in patients with myelodysplastic syndromes. Cancer. 2002; 95: 1923–30.
64. de Bont ES, Fidler V, Meeuwsen T, Scherpen F, Hahlen K, Kamps WA. Vascular endothelial growth factor secretion is an independent prognostic factor for relapse-free survival in pediatric acute myeloid leukemia patients. Clin Cancer Res. 2002; 8: 2856–61.
65. Rosen LS. Clinical experience with angiogenesis signaling inhibitors: focus on vascular endothelial growth factor (VEGF) blockers. Cancer Control. 2002; 9(2 Suppl): 36–44.
66. Folkman J. Role of angiogenesis in tumor growth and metastasis. Semin Oncol. 2002; 29(6 Suppl 16): 15–18.
67. Tischer E, Mitchell R, Hartman T, Silva M, Gospodarowicz D, Fiddes JC, Abraham JA. The human gene for vascular endothelial growth factor: multiple protein forms are encoded through alternative exon splicing. J Biol Chem. 1991; 266: 11947–54.
68. Lander HM, Jacovina AT, Davis RJ, Tauras JM. Differential activation of mitogen-activated protein kinases by nitric oxide-related species. J Biol Chem. 1996; 271: 19705–09.
69. Benckert C, Jonas S, Cramer T, Von Marschall Z, Schafer G, Peters M, Wagner K, Radke C, Wiedenmann B, Neuhaus P, Hocker M, Rosewicz S. Transforming growth factor beta 1 stimulates vascular endothelial growth factor gene transcription in human cholangiocellular carcinoma cells. Cancer Res. 2003; 63: 1083–92.
70. Shi Q, Le X, Abbruzzese JL, Peng Z, Qian CN, Tang H, Xiong Q, Wang B, Li XC, Xie K. Constitutive Sp1 activity is essential for differential constitutive expression of vascular endothelial growth factor in human pancreatic adenocarcinoma. Cancer Res. 2001; 61: 4143–54.

71. Sanchez-Elsner T, Botella LM, Velasco B, Corbi A, Attisano L, Bernabeu C. Synergistic cooperation between hypoxia and transforming growth factor-beta pathways on human vascular endothelial growth factor gene expression. J Biol Chem. 2001; 276: 38527–35.
72. Kimura H, Weisz A, Kurashima Y, Hashimoto K, Ogura T, D'Acquisto F, Addeo R, Makuuchi M, Esumi H. Hypoxia response element of the human vascular endothelial growth factor gene mediates transcriptional regulation by nitric oxide: control of hypoxia-inducible factor-1 activity by nitric oxide. Blood. 2000; 95: 189–97.
73. Wei D, Le X, Zheng L, Wang L, Frey JA, Gao AC, Peng Z, Huang S, Xiong HQ, Abbruzzese JL, Xie K. Stat3 activation regulates the expression of vascular endothelial growth factor and human pancreatic cancer angiogenesis and metastasis. Oncogene. 2003; 22: 319–29.
74. Mueller MD, Vigne JL, Minchenko A, Lebovic DI, Leitman DC, Taylor RN. Regulation of vascular endothelial growth factor (VEGF) gene transcription by estrogen receptors alpha and beta. Proc Natl Acad Sci USA. 2000; 97: 10972–77.
75. Kadonaga JT, Carner KR, Masiarz FR, Tjian R. Isolation of cDNA encoding transcription factor Sp1 and functional analysis of the DNA binding domain. Cell. 1987; 51: 1079–90.
76. Reisinger K, Kaufmann R, Gille J. Increased Sp1 phosphorylation as a mechanism of hepatocyte growth factor (HGF/SF)-induced vascular endothelial growth factor (VEGF/VPF) transcription. J Cell Sci. 2003; 116(Pt 2): 225–38.
77. Ryuto M, Ono M, Izumi H, Yoshida S, Weich HA, Kohno K, Kuwano M. Induction of vascular endothelial growth factor by tumor necrosis factor alpha in human glioma cells. Possible roles of SP-1. J Biol Chem. 1996; 271: 28220–28.
78. Siemeister G, Weindel K, Mohrs K, Barleon B, Martiny-Baron G, Marme D. Reversion of deregulated expression of vascular endothelial growth factor in human renal carcinoma cells by von Hippel-Lindau tumor suppressor protein. Cancer Res. 1996; 56: 2299–301.
79. Mukhopadhyay D, Knebelmann B, Cohen HT, Ananth S, Sukhatme VP. The von Hippel-Lindau tumor suppressor gene product interacts with Sp1 to repress vascular endothelial growth factor promoter activity. Mol Cell Biol. 1997; 17: 5629–39.
80. Cohen HT, Zhou M, Welsh AM, Zarghamee S, Scholz H, Mukhopadhyay D, Kishida T, Zbar B, Knebelmann B, Sukhatme VP. An important von Hippel-Lindau tumor suppressor domain mediates Sp1-binding and self-association. Biochem Biophys Res Commun. 1999; 266: 43–50.
81. Zhang L, Yu D, Hu M, Xiong S, Lang A, Ellis LM, Pollock RE. Wild-type p53 suppresses angiogenesis in human leiomyosarcoma and synovial sarcoma by transcriptional suppression of vascular endothelial growth factor expression. Cancer Res. 2000; 60: 3655–61.
82. Pal S, Datta K, Mukhopadhyay D. Central role of p53 on regulation of vascular permeability factor/vascular endothelial growth factor (VPF/VEGF) expression in mammary carcinoma. Cancer Res. 2001; 61: 6952–57.
83. Salimath B, Marme D, Finkenzeller G. Expression of the vascular endothelial growth factor gene is inhibited by p73. Oncogene. 2000; 19: 3470–76.
84. Mukhopadhyay D, Tsiokas L, Zhou XM, Foster D, Brugge JS, Sukhatme VP. Hypoxic induction of human vascular endothelial growth factor expression through c-Src activation. Nature. 1995; 375(6532): 577–81.
85. Rak J, Filmus J, Finkenzeller G, Grugel S, Marme D, Kerbel RS. Oncogenes as inducers of tumor angiogenesis. Cancer Metastasis Rev. 1995; 14: 263–77.
86. Schafer G, Cramer T, Suske G, Kemmner W, Wiedenmann B, Hocker M. Oxidative stress regulates vascular endothelial growth factor-A gene transcription through Sp1- and Sp3-dependent activation of two proximal GC-rich promoter elements. J Biol Chem. 2003; 278: 8190–98.
87. Berra E, Milanini J, Richard DE, Le Gall M, Vinals F, Gothie E, Roux D, Pages G, Pouyssegur J. Signaling angiogenesis via p42/p44 MAP kinase and hypoxia. Biochem Pharmacol. 2000; 60: 1171–78.

88. Laughner E, Taghavi P, Chiles K, Mahon PC, Semenza GL. HER2 (neu) signaling increases the rate of hypoxia-inducible factor 1alpha (HIF-1alpha) synthesis: novel mechanism for HIF-1-mediated vascular endothelial growth factor expression. Mol Cell Biol. 2001; 21: 3995–4004.
89. Minet E, Arnould T, Michel G, Roland I, Mottet D, Raes M, Remacle J, Michiels C. ERK activation upon hypoxia: involvement in HIF-1 activation. FEBS Lett. 2000; 468: 53–58.
90. Wei LH, Kuo ML, Chen CA, Chou CH, Lai KB, Lee CN, Hsieh CY. Interleukin-6 promotes cervical tumor growth by VEGF-dependent angiogenesis via a STAT3 pathway. Oncogene. 2003; 22: 1517–27.
91. Niu G, Wright KL, Huang M, Song L, Haura E, Turkson J, Zhang S, Wang T, Sinibaldi D, Coppola D, Heller R, Ellis LM, Karras J, Bromberg J, Pardoll D, Jove R, Yu H. Constitutive Stat3 activity up-regulates VEGF expression and tumor angiogenesis. Oncogene. 2002; 21: 2000–2008.
92. Funamoto M, Fujio Y, Kunisada K, Negoro S, Tone E, Osugi T, Hirota H, Izumi M, Yoshizaki K, Walsh K, Kishimoto T, Yamauchi-Takihara K. Signal transducer and activator of transcription 3 is required for glycoprotein 130-mediated induction of vascular endothelial growth factor in cardiac myocytes. J Biol Chem. 2000; 275: 10561–66.
93. Buettner R, Mora LB, Jove R. Activated STAT signaling in human tumors provides novel molecular targets for therapeutic intervention. Clin Cancer Res. 2002; 8: 945–54.
94. Leonard WJ. Role of Jak kinases and STATs in cytokine signal transduction. Int J Hematol. 2001; 73: 271–77.
95. Turkson J, Jove R. STAT proteins: novel molecular targets for cancer drug discovery. Oncogene. 2000; 19: 6613–26.
96. Yahata Y, Shirakata Y, Tokumaru S, Yamasaki K, Sayama K, Hanakawa Y, Detmar M, Hashimoto K. Nuclear translocation of phosphorylated STAT3 is essential for vascular endothelial growth factor-induced human dermal microvascular endothelial cell migration and tube formation. J Biol Chem. 2003; 278: 40026–31.
97. Swannie HC, Kaye SB. Protein kinase C inhibitors. Curr Oncol Rep. 2002; 4: 37–46.
98. Neri LM, Borgatti P, Capitani S, Martelli AM. Protein kinase C isoforms and lipid second messengers: a critical nuclear partnership? Histol Histopathol. 2002; 17: 1311–16.
99. Nishizuka Y. Intracellular signaling by hydrolysis of phospholipids and activation of protein kinase C. Science. 1992; 258: 607–14.
100. Michiels C, Minet E, Michel G, Mottet D, Piret JP, Raes M. HIF-1 and AP-1 cooperate to increase gene expression in hypoxia: role of MAP kinases. IUBMB Life. 2001; 52: 49–53.
101. Karin M, Shaulian E. AP-1: linking hydrogen peroxide and oxidative stress to the control of cell proliferation and death. IUBMB Life. 2001; 52: 17–24.
102. Tsai JC, Teng LJ, Chen CT, Hong TM, Goldman CK, Gillespie GY. Protein kinase C mediates induced secretion of vascular endothelial growth factor by human glioma cells. Biochem Biophys Res Commun. 2003; 309: 952–60.
103. Minet E, Michel G, Mottet D, Piret JP, Barbieux A, Raes M, Michiels C. c-JUN gene induction and AP-1 activity is regulated by a JNK-dependent pathway in hypoxic HepG2 cells. Exp Cell Res. 2001; 265: 114–24.
104. Karin M, Delhase M. JNK or IKK, AP-1 or NF-kappaB, which are the targets for MEK kinase 1 action? Proc Natl Acad Sci USA. 1998; 95: 9067–69.
105. Karin M, Liu Z, Zandi E. AP-1 function and regulation. Curr Opin Cell Biol. 1997; 9: 240–46.
106. Gille J, Swerlick RA, Caughman SW. Transforming growth factor-alpha-induced transcriptional ctivation of the vascular permeability factor (VPF/VEGF) gene requires AP-2-dependent DNA binding and transactivation. EMBO J. 1997; 16: 750–59.
107. Brenneisen P, Blaudschun R, Gille J, Schneider L, Hinrichs R, Wlaschek M, Eming S, Scharffetter-Kochanek K. Essential role of an activator protein-2 (AP-2)/specificity protein 1 (Sp1) cluster in the UVB-mediated induction of the human vascular endothelial growth factor in HaCaT keratinocytes. Biochem J. 2003; 369(Pt 2): 341–49.

108. Schwarte-Waldhoff I, Schmiegel W. Smad4 transcriptional pathways and angiogenesis. Int J Gastrointest Cancer. 2002; 31: 47–59.
109. Schwarte-Waldhoff I, Volpert OV, Bouck NP, Sipos B, Hahn SA, Klein-Scory S, Luttges J, Kloppel G, Graeven U, Eilert-Micus C, Hintelmann A, Schmiegel W. Smad4/DPC4-mediated tumor suppression through suppression of angiogenesis. Proc Natl Acad Sci USA. 2000; 97: 9624–29.
110. Damert A, Ikeda E, Risau W. Activator-protein-1 binding potentiates the hypoxia-induciblefactor-1-mediated hypoxia-induced transcriptional activation of vascular-endothelial growth factor expression in C6 glioma cells. Biochem J. 1997; 327(Pt 2): 419–23.
111. Milanini J, Vinals F, Pouyssegur J, Pages G. p42/p44 MAP kinase module plays a key role in the transcriptional regulation of the vascular endothelial growth factor gene in fibroblasts. J Biol Chem. 1998; 273: 18165–72.
112. Blaudschun R, Sunderkotter C, Brenneisen P, Hinrichs R, Peters T, Schneider L, Razi-Wolf Z, Hunzelmann N, Scharffetter-Kochanek K. Vascular endothelial growth factor causally contributes to the angiogenic response upon ultraviolet B irradiation in vivo. Br J Dermatol. 2002; 146: 581–87.
113. Levy AP, Levy NS, Goldberg MA. Post-transcriptional regulation of vascular endothelial growth factor by hypoxia. J Biol Chem. 1996; 271: 2746–53.
114. Levy NS, Chung S, Furneaux H, Levy AP. Hypoxic stabilization of vascular endothelial growth factor mRNA by the RNA-binding protein HuR. J Biol Chem. 1998; 273: 6417–23.
115. Nabors LB, Gillespie GY, Harkins L, King PH. HuR, a RNA stability factor, is expressed in malignant brain tumors and binds to adenine- and uridine-rich elements within the 3' untranslated regions of cytokine and angiogenic factor mRNAs. Cancer Res. 2001; 61: 2154–61.
116. Dibbens JA, Miller DL, Damert A, Risau W, Vadas MA, Goodall GJ. Hypoxic regulation of vascular endothelial growth factor mRNA stability requires the cooperation of multiple RNA elements. Mol Biol Cell. 1999; 10: 907–19.
117. Pages G, Berra E, Milanini J, Levy AP, Pouyssegur J. Stress-activated protein kinases (JNK and p38/HOG) are essential for vascular endothelial growth factor mRNA stability. J Biol Chem. 2000; 275: 26484–91.
118. Prats AC, Prats H. Translational control of gene expression: role of IRESs and consequences for cell transformation and angiogenesis. Prog Nucleic Acid Res Mol Biol. 2002; 72: 367–413.
119. Guo P, Xu L, Pan S, Brekken RA, Yang ST, Whitaker GB, Nagane M, Thorpe PE, Rosenbaum JS, Su Huang HJ, Cavenee WK, Cheng SY. Vascular endothelial growth factor isoforms display distinct activities in promoting tumor angiogenesis at different anatomic sites. Cancer Res. 2001; 61: 8569–77.
120. Stein I, Itin A, Einat P, Skaliter R, Grossman Z, Keshet E. Translation of vascular endothelial growth factor mRNA by internal ribosome entry: implications for translation under hypoxia. Mol Cell Biol. 1998; 18: 3112–19.
121. Huez I, Creancier L, Audigier S, Gensac MC, Prats AC, Prats H. Two independent internal ribosome entry sites are involved in translation initiation of vascular endothelial growth factor mRNA. Mol Cell Biol. 1998; 18: 6178–90.
122. Kevil CG, De Benedetti A, Payne DK, Coe LL, Laroux FS, Alexander JS. Translational regulation of vascular permeability factor by eukaryotic initiation factor 4E: implications for tumor angiogenesis. Int J Cancer. 1996; 65: 785–90.
123. Li BD, Liu L, Dawson M, De Benedetti A. Overexpression of eukaryotic initiation factor 4E (eIF4E) in breast carcinoma. Cancer. 1997; 79: 2385–90.
124. Crew JP, Fuggle S, Bicknell R, Cranston DW, de Benedetti A, Harris AL. Eukaryotic initiation factor-4E in superficial and muscle invasive bladder cancer and its correlation with vascular endothelial growth factor expression and tumour progression. Br J Cancer. 2000; 82: 161–66.
125. Kuwabara K, Matsumoto M, Ikeda J, Hori O, Ogawa S, Maeda Y, Kitagawa K, Imuta N, Kinoshita T, Stern DM, Yanagi H, Kamada T. Purification and characterization of a novel

stress protein, the 150-kDa oxygen-regulated protein (ORP150), from cultured rat astrocytes and its expression in ischemic mouse brain. J Biol Chem. 1996; 271: 5025–32.
126. Ozawa K, Kondo T, Hori O, Kitao Y, Stern DM, Eisenmenger W, Ogawa S, Ohshima T. Expression of the oxygen-regulated protein ORP150 accelerates wound healing by modulating intracellular VEGF transport. J Clin Invest. 2001; 108: 41–50.
127. Miyagi T, Hori O, Koshida K, Egawa M, Kato H, Kitagawa Y, Ozawa K, Ogawa S, Namiki M. Antitumor effect of reduction of 150-kDa oxygen-regulated protein expression on human prostate cancer cells. Int J Urol. 2002; 9: 577–85.
128. Carmeliet P, Jain RK. Angiogenesis in cancer and other diseases. Nature. 2000; 407(6801): 249–57.
129. Igarashi H, Esumi M, Ishida H, Okada K. Vascular endothelial growth factor overexpression is correlated with von Hippel-Lindau tumor suppressor gene inactivation in patients with sporadic renal cell carcinoma. Cancer. 2002; 95: 47–53.
130. Okada F, Rak JW, Croix BS, Lieubeau B, Kaya M, Roncari L, Shirasawa S, Sasazuki T, Kerbel RS. Impact of oncogenes in tumor angiogenesis: mutant K-ras up-regulation of vascular endothelial growth factor/vascular permeability factor is necessary, but not sufficient for tumorigenicity of human colorectal carcinoma cells. Proc Natl Acad Sci USA. 1998; 95: 3609–14.
131. Rak J, Mitsuhashi Y, Bayko L, Filmus J, Shirasawa S, Sasazuki T, Kerbel RS. Mutant ras oncogenes upregulate VEGF/VPF expression: implications for induction and inhibition of tumor angiogenesis. Cancer Res. 1995; 55: 4575–80.
132. Larcher F, Robles AI, Duran H, Murillas R, Quintanilla M, Cano A, Conti CJ, Jorcano JL. Up-regulation of vascular endothelial growth factor/vascular permeability factor in mouse skin carcinogenesis correlates with malignant progression state and activated H-ras expression levels. Cancer Res. 1996; 56: 5391–96.
133. D'Arcangelo G, Habas R, Wang S, Halegoua S, Salton SR. Activation of codependent transcription factors is required for transcriptional induction of the vgf gene by nerve growth factor and Ras. Mol Cell Biol. 1996; 16: 4621–31.
134. White FC, Benehacene A, Scheele JS, Kamps M. VEGF mRNA is stabilized by ras and tyrosine kinase oncogenes, as well as by UV radiation – evidence for divergent stabilization pathways. Growth Factors. 1997; 14: 199–212.
135. Sodhi A, Montaner S, Miyazaki H, Gutkind JS. MAPK and Akt act cooperatively but independently on hypoxia inducible factor-1alpha in rasV12 upregulation of VEGF. Biochem Biophys Res Commun. 2001; 287: 292–300.
136. Milanini-Mongiat J, Pouyssegur J, Pages G. Identification of two Sp1 phosphorylation sites for p42/p44 mitogen-activated protein kinases: their implication in vascular endothelial growth factor gene transcription. J Biol Chem. 2002; 277: 20631–39.
137. Pal S, Datta K, Khosravi-Far R, Mukhopadhyay D. Role of protein kinase C-zeta in Ras-mediated transcriptional activation of vascular permeability factor/vascular endothelial growth factor expression. J Biol Chem. 2001; 276: 2395–403.
138. Pal S, Claffey KP, Cohen HT, Mukhopadhyay D. Activation of Sp1-mediated vascular permeability factor/vascular endothelial growth factor transcription requires specific interaction with protein kinase C-zeta. J Biol Chem. 1998; 273: 26277–80.
139. Hoshino R, Chatani Y, Yamori T, Tsuruo T, Oka H, Yoshida O, Shimada Y, Ari-i S, Wada H, Fujimoto J, Kohno M. Constitutive activation of the 41-/43-kDa mitogen-activated protein kinase signaling pathway in human tumors. Oncogene. 1999; 18: 813–22.
140. Salh B, Marotta A, Matthewson C, Ahluwalia M, Flint J, Owen D, Pelech S. Investigation of the Mek-MAP kinase-Rsk pathway in human breast cancer. Anticancer Res. 1999; 19(1B): 731–40.
141. Schmidt CM, McKillop IH, Cahill PA, Sitzmann JV. Increased MAPK expression and activity in primary human hepatocellular carcinoma. Biochem Biophys Res Commun. 1997; 236: 54–58.
142. Govindarajan B, Bai X, Cohen C, Zhong H, Kilroy S, Louis G, Moses M, Arbiser JL. Malignant transformation of melanocytes to melanoma by constitutive activation of

mitogen-activated protein kinase kinase (MAPKK) signaling. J Biol Chem. 2003; 278: 9790–95.
143. Cohen C, Zavala-Pompa A, Sequeira JH, Shoji M, Sexton DG, Cotsonis G, Cerimele F, Govindarajan B, Macaron N, Arbiser JL. Mitogen-actived protein kinase activation is an early event in melanoma progression. Clin Cancer Res. 2002; 8: 3728–33.
144. Katsanakis KD, Owen C, Zoumpourlis V. JNK and ERK signaling pathways in multistage mouse carcinogenesis: studies in the inhibition of signaling cascades as a means to understand their in vivo biological role. Anticancer Res. 2002; 22: 755–59.
145. Mazure NM, Chen EY, Laderoute KR, Giaccia AJ. Induction of vascular endothelial growth factor by hypoxia is modulated by a phosphatidylinositol 3-kinase/Akt signaling pathway in Ha-ras-transformed cells through a hypoxia inducible factor-1 transcriptional element. Blood. 1997; 90: 3322–31.
146. Richard DE, Berra E, Gothie E, Roux D, Pouyssegur J. p42/p44 mitogen-activated protein kinases phosphorylate hypoxia-inducible factor 1alpha (HIF-1alpha) and enhance the transcriptional activity of HIF-1. J Biol Chem. 1999; 274: 32631–37.
147. Urata Y, Yamaguchi M, Higashiyama Y, Ihara Y, Goto S, Kuwano M, Horiuchi S, Sumikawa K, Kondo T. Reactive oxygen species accelerate production of vascular endothelial growth factor by advanced glycation end products in RAW264.7 mouse macrophages. Free Radic Biol Med. 2002; 32(8): 688–701.
148. Okajima E, Thorgeirsson UP. Different regulation of vascular endothelial growth factor expression by the ERK and p38 kinase pathways in v-ras, v-raf, and v-myc transformed cells. Biochem Biophys Res Commun. 2000; 270: 108–11.
149. Miao H, Wei BR, Peehl DM, Li Q, Alexandrou T, Schelling JR, Rhim JS, Sedor JR, Burnett E, Wang B. Activation of EphA receptor tyrosine kinase inhibits the Ras/MAPK pathway. Nat Cell Biol. 2001; 3: 527–30.
150. Zhang X, Gaspard JP, Chung DC. Regulation of vascular endothelial growth factor by the Wnt and K-ras pathways in colonic neoplasia. Cancer Res. 2001; 61: 6050–54.
151. Sheta EA, Harding MA, Conaway MR, Theodorescu D. Focal adhesion kinase, Rap1, and transcriptional induction of vascular endothelial growth factor. J Natl Cancer Inst. 2000; 92: 1065–73.
152. Thomas SM, Brugge JS. Cellular functions regulated by Src family kinases. Annu Rev Cell Dev Biol. 1997; 13: 513–609.
153. Gruden G, Thomas S, Burt D, Lane S, Chusney G, Sacks S, Viberti G. Mechanical stretch induces vascular permeability factor in human mesangial cells: mechanisms of signal transduction. Proc Natl Acad Sci USA. 1997; 94: 12112–16.
154. Gruden G, Araf S, Zonca S, Burt D, Thomas S, Gnudi L, Viberti G. IGF-I induces vascular endothelial growth factor in human mesangial cells via a Src-dependent mechanism. Kidney Int. 2003; 63: 1249–55.
155. Mukhopadhyay D, Tsiokas L, Sukhatme VP. Wild-type p53 and v-Src exert opposing influences on human vascular endothelial growth factor gene expression. Cancer Res. 1995; 55: 6161–65.
156. Fleming RY, Ellis LM, Parikh NU, Liu W, Staley CA, Gallick GE. Regulation of vascular endothelial growth factor expression in human colon carcinoma cells by activity of src kinase. Surgery. 1997; 122: 501–07.
157. Ellis LM, Staley CA, Liu W, Fleming RY, Parikh NU, Bucana CD, Gallick GE. Down-regulation of vascular endothelial growth factor in a human colon carcinoma cell line transfected with an antisense expression vector specific for c-src. J Biol Chem. 1998; 273: 1052–57.
158. Theurillat JP, Hainfellner J, Maddalena A, Weissenberger J, Aguzzi A. Early induction of angiogenetic signals in gliomas of GFAP-v-src transgenic mice. Am J Pathol. 1999; 154: 581–90.
159. Jiang BH, Agani F, Passaniti A, Semenza GL. V-SRC induces expression of hypoxia-inducible factor 1 (HIF-1) and transcription of genes encoding vascular endothelial growth factor and enolase 1: involvement of HIF-1 in tumor progression. Cancer Res. 1997; 57: 5328–35.

160. Lee SW, Kim WJ, Choi YK, Song HS, Son MJ, Gelman IH, Kim YJ, Kim KW. SSeCKS regulates angiogenesis and tight junction formation in blood-brain barrier. Nat Med. 2003; 9: 900–06.
161. Kumar R, Yarmand-Bagheri R. The role of HER2 in angiogenesis. Semin Oncol. 2001; 28(5 Suppl 16): 27–32.
162. Petit AM, Rak J, Hung MC, Rockwell P, Goldstein N, Fendly B, Kerbel RS. Neutralizing antibodies against epidermal growth factor and ErbB-2/neu receptor tyrosine kinases down-regulate vascular ndothelial growth factor production by tumor cells in vitro and in vivo: angiogenic implications for signal transduction therapy of solid tumors. Am J Pathol. 1997; 151: 1523–30.
163. Koukourakis MI, Giatromanolaki A, O'Byrne KJ, Cox J, Krammer B, Gatter KC, Harris AL. bcl-2 and c-erbB-2 proteins are involved in the regulation of VEGF and of thymidine phosphorylase angiogenic activity in non-small-cell lung cancer. Clin Exp Metastasis. 1999; 17: 545–54.
164. Anan K, Morisaki T, Katano M, Ikubo A, Tsukahara Y, Kojima M, Uchiyama A, Kuroki S, Torisu M, Tanaka M. Assessment of c-erbB2 and vascular endothelial growth factor mRNA expression in fine-needle aspirates from early breast carcinomas: pre-operative determination of malignant potential. Eur J Surg Oncol. 1998; 24: 28–33.
165. Yang W, Klos K, Yang Y, Smith TL, Shi D, Yu D. ErbB2 overexpression correlates with increased expression of vascular endothelial growth factors A, C, and D in human breast carcinoma. Cancer. 2002; 94: 2855–61.
166. Lee JS, Kim HS, Jung JJ, Kim YB, Lee MC, Park CS. Expression of vascular endothelial growth factor in invasive ductal carcinoma of the breast and the relation to angiogenesis and p53 and HER-2/neu protein expression. Appl Immunohistochem Mol Morphol. 2002; 10: 289–95.
167. Lee JS, Kim HS, Jung JJ, Lee MC, Park CS. Expression of vascular endothelial growth factor in adenocarcinomas of the uterine cervix and its relation to angiogenesis and p53 and c-erbB-2 protein expression. Gynecol Oncol. 2002; 85: 469–75.
168. Yen L, Benlimame N, Nie ZR, Xiao D, Wang T, Al Moustafa AE, Esumi H, Milanini J, Hynes NE, Pages G, Alaoui-Jamali MA. Differential regulation of tumor angiogenesis by distinct ErbB homo- and heterodimers. Mol Biol Cell. 2002; 13: 4029–44.
169. Mayerhofer M, Valent P, Sperr WR, Griffin JD, Sillaber C. BCR/ABL induces expression of vascular endothelial growth factor and its transcriptional activator, hypoxia inducible factor-1alpha, through a pathway involving phosphoinositide 3-kinase and the mammalian target of rapamycin. Blood. 2002; 100: 3767–75.
170. Ebos JM, Tran J, Master Z, Dumont D, Melo JV, Buchdunger E, Kerbel RS. Imatinib mesylate (STI-571) reduces Bcr-Abl-mediated vascular endothelial growth factor secretion in chronic myelogenous leukemia. Mol Cancer Res. 2002; 1: 89–95.
171. Lopez-Ocejo O, Viloria-Petit A, Bequet-Romero M, Mukhopadhyay D, Rak J, Kerbel RS. Oncogenes and tumor angiogenesis: the HPV-16 E6 oncoprotein activates the vascular endothelial growth factor (VEGF) gene promoter in a p53 independent manner. Oncogene. 2000; 19: 4611–20.
172. Fontanini G, Boldrini L, Vignati S, Chine S, Basolo F, Silvestri V, Lucchi M, Mussi A, Angeletti CA, Bevilacqua G. Bcl2 and p53 regulate vascular endothelial growth factor (VEGF)-mediated angiogenesis in non-small cell lung carcinoma. Eur J Cancer. 1998; 34: 718–23.
173. Kobayashi N, Ishii M, Ueno Y, Kisara N, Chida N, Iwasaki T, Toyota T. Co-expression of Bcl-2 protein and vascular endothelial growth factor in hepatocellular carcinomas treated by hemoembolization. Liver. 1999; 19: 25–31.
174. Biroccio A, Candiloro A, Mottolese M, Sapora O, Albini A, Zupi G, Del Bufalo D. Bcl-2 overexpression and hypoxia synergistically act to modulate vascular endothelial growth factor expression and in vivo angiogenesis in a breast carcinoma line. FASEB J. 2000; 14: 652–60.
175. Wizigmann-Voos S, Breier G, Risau W, Plate KH. Up-regulation of vascular endothelial growth factor and its receptors in von Hippel-Lindau disease-associated and sporadic hemangioblastomas. Cancer Res. 1995; 55: 1358–64.

176. Siemeister G, Weindel K, Mohrs K, Barleon B, Martiny-Baron G, Marme D. Reversion of deregulated expression of vascular endothelial growth factor in human renal carcinoma cells by von Hippel-Lindau tumor suppressor protein. Cancer Res. 1996; 56: 2299–301.
177. Ohh M, Takagi Y, Aso T, Stebbins CE, Pavletich NP, Zbar B, Conaway RC, Conaway JW, Kaelin WG Jr. Synthetic peptides define critical contacts between elongin C, elongin B, and the von Hippel-Lindau protein. J Clin Invest. 1999; 104: 1583–91.
178. Gnarra JR, Zhou S, Merrill MJ, Wagner JR, Krumm A, Papavassiliou E, Oldfield EH, Klausner RD, Linehan WM. Post-transcriptional regulation of vascular endothelial growth factor mRNA by the product of the VHL tumor suppressor gene. Proc Natl Acad Sci USA. 1996; 93: 10589–94.
179. Iliopoulos O, Levy AP, Jiang C, Kaelin WG Jr, Goldberg MA. Negative regulation of hypoxia-inducible genes by the von Hippel-Lindau protein. Proc Natl Acad Sci USA. 1996; 93: 10595–99.
180. Levy AP, Levy NS, Goldberg MA. Hypoxia-inducible protein binding to vascular endothelial growth factor mRNA and its modulation by the von Hippel-Lindau protein. J Biol Chem. 1996; 271: 25492–97.
181. Pal S, Claffey KP, Dvorak HF, Mukhopadhyay D. The von Hippel-Lindau gene product inhibits vascular permeability factor/vascular endothelial growth factor expression in renal cell carcinoma by blocking protein kinase C pathways. J Biol Chem. 1997; 272: 27509–12.
182. Pioli PA, Rigby WF. The von Hippel-Lindau protein interacts with heteronuclear ribonucleoprotein a2 and regulates its expression. J Biol Chem. 2001; 276: 40346–52.
183. Li Z, Wang D, Na X, Schoen SR, Messing EM, Wu G. The VHL protein recruits a novel KRAB-A domain protein to repress HIF-1alpha transcriptional activity. EMBO J. 2003; 22: 1857–67.
184. Kieser A, Weich HA, Brandner G, Marme D, Kolch W. Mutant p53 potentiates protein kinase C induction of vascular endothelial growth factor expression. Oncogene. 1994; 9: 963–69.
185. Bouvet M, Ellis LM, Nishizaki M, Fujiwara T, Liu W, Bucana CD, Fang B, Lee JJ, Roth JA. Adenovirus-mediated wild-type p53 gene transfer down-regulates vascular endothelial growth factor expression and inhibits angiogenesis in human colon cancer. Cancer Res. 1998; 58: 2288–92.
186. Choi KS, Bae MK, Jeong JW, Moon HE, Kim KW. Hypoxia-induced angiogenesis during carcinogenesis. J Biochem Mol Biol. 2003; 36: 120–27.
187. Ravi R, Mookerjee B, Bhujwalla ZM, Sutter CH, Artemov D, Zeng Q, Dillehay LE, Madan A, Semenza GL, Bedi A. Regulation of tumor angiogenesis by p53-induced degradation of hypoxia-inducible factor 1alpha. Genes Dev. 2000; 14: 34–44.
188. Bae MK, Ahn MY, Jeong JW, Bae MH, Lee YM, Bae SK, Park JW, Kim KR, Kim KW. Jab1 interacts directly with HIF-1alpha and regulates its stability. J Biol Chem. 2002; 277: 9–12.
189. Pugh CW, Ratcliffe PJ. The von Hippel-Lindau tumor suppressor, hypoxia-inducible factor-1 (HIF-1) degradation, and cancer pathogenesis. Semin Cancer Biol. 2003; 13: 83–89.
190. Kondo K, Klco J, Nakamura E, Lechpammer M, Kaelin WG Jr. Inhibition of HIF is necessary for tumor suppression by the von Hippel-Lindau protein. Cancer Cell. 2002; 1: 237–46.
191. Fontanini G, Vignati S, Lucchi M, Mussi A, Calcinai A, Boldrini L, Chine S, Silvestri V, Angeletti CA, Basolo F, Bevilacqua G. Neoangiogenesis and p53 protein in lung cancer: their prognostic role and their relation with vascular endothelial growth factor (VEGF) expression. Br J Cancer. 1997; 75: 1295–301.
192. Kang SM, Maeda K, Onoda N, Chung YS, Nakata B, Nishiguchi Y, Sowa M. Combined analysis of p53 and vascular endothelial growth factor expression in colorectal carcinoma for determination of tumor vascularity and liver metastasis. Int J Cancer. 1997; 74: 502–07.
193. Takahashi Y, Bucana CD, Cleary KR, Ellis LM. p53, vessel count, and vascular endothelial growth factor expression in human colon cancer. Int J Cancer. 1998; 79: 34–38.
194. Kondo Y, Arii S, Furutani M, Isigami S, Mori A, Onodera H, Chiba T, Imamura M. Implication of vascular endothelial growth factor and p53 status for angiogenesis in noninvasive colorectal carcinoma. Cancer. 2000; 88: 1820–27.

195. Yuan A, Yu CJ, Luh KT, Kuo SH, Lee YC, Yang PC. Aberrant p53 expression correlates with expression of vascular endothelial growth factor mRNA and interleukin-8 mRNA and neoangiogenesis in non-small-cell lung cancer. J Clin Oncol. 2002; 20: 900–10.
196. Uchida S, Shimada Y, Watanabe G, Tanaka H, Shibagaki I, Miyahara T, Ishigami S, Imamura M. In oesophageal squamous cell carcinoma vascular endothelial growth factor is associated with p53 mutation, advanced stage and poor prognosis. Br J Cancer. 1998; 77: 1704–09.
197. Zietz C, Rossle M, Haas C, Sendelhofert A, Hirschmann A, Sturzl M, Lohrs U. MDM-2 oncoprotein overexpression, p53 gene mutation, and VEGF up-regulation in angiosarcomas. Am J Pathol. 1998; 153: 1425–33.
198. Linderholm BK, Lindahl T, Holmberg L, Klaar S, Lennerstrand J, Henriksson R, Bergh J. The expression of vascular endothelial growth factor correlates with mutant p53 and poor prognosis in human breast cancer. Cancer Res. 2001; 61: 2256–60.
199. Saito H, Tujitani S, Ikeguchi M, Maeta M, Kaibara N. Neoangiogenesis and relationship to nuclear p53 accumulation and vascular endothelial growth factor expression in advanced gastric carcinoma. Oncology. 1999; 57: 164–72.
200. Giatromanolaki A, Koukourakis MI, Kakolyris S, Turley H, O'Byrne K, Scott PA, Pezzella F, Georgoulias V, Harris AL, Gatter KC. Vascular endothelial growth factor, wild-type p53, and angiogenesis in early operable non-small cell lung cancer. Clin Cancer Res. 1998; 4: 3017–24.
201. Ambs S, Bennett WP, Merriam WG, Ogunfusika MO, Oser SM, Khan MA, Jones RT, Harris CC. Vascular endothelial growth factor and nitric oxide synthase expression in human lung cancer and the elation to p53. Br J Cancer. 1998; 78: 233–39.
202. Maeda T, Matsumura S, Hiranuma H, Jikko A, Furukawa S, Ishida T, Fuchihata H. Expression of vascular endothelial growth factor in human oral squamous cell carcinoma: its association with tumour progression and p53 gene status. J Clin Pathol. 1998; 51: 771–75.
203. Rosa AR, Schirmer CC, Gurski RR, Meurer L, Edelweiss MI, Kruel CD. Prognostic value of p53 protein expression and vascular endothelial growth factor expression in resected squamous cell carcinoma of the esophagus. Dis Esophagus. 2003; 16: 112–18.
204. Cascinu S, Graziano F, Catalano V, Barni S, Giordani P, Baldelli AM, Staccioli MP, Rossi C, Brenna A, Valenti A, Muretto P, Catalano G. Vascular endothelial growth factor and p53 expressions in liver and abdominal metastases from colon cancer. Tumour Biol. 2003; 24: 77–81.
205. Agani F, Kirsch DG, Friedman SL, Kastan MB, Semenza GL. p53 does not repress hypoxia-induced transcription of the vascular endothelial growth factor gene. Cancer Res. 1997; 57: 4474–77.
206. Vikhanskaya F, Bani MR, Borsotti P, Ghilardi C, Ceruti R, Ghisleni G, Marabese M, Giavazzi R, Broggini M, Taraboletti G. p73 overexpression increases VEGF and reduces thrombospondin-1 production: implications for tumor angiogenesis. Oncogene. 2001; 20: 7293–300.
207. Tsugawa K, Jones MK, Sugimachi K, Sarfeh IJ, Tarnawski AS. Biological role of phosphatase PTEN in cancer and tissue injury healing. Front Biosci. 2002; 7: e245–51.
208. Zundel W, Schindler C, Haas-Kogan D, Koong A, Kaper F, Chen E, Gottschalk AR, Ryan HE, Johnson RS, Jefferson AB, Stokoe D, Giaccia AJ. Loss of PTEN facilitates HIF-1-mediated gene expression. Genes Dev. 2000; 14: 391–96.
209. Zhong H, Chiles K, Feldser D, Laughner E, Hanrahan C, Georgescu MM, Simons JW, Semenza GL. Modulation of hypoxia-inducible factor 1alpha expression by the epidermal growth actor/phosphatidylinositol 3-kinase/PTEN/AKT/FRAP pathway in human prostate cancer cells: implications for tumor angiogenesis and therapeutics. Cancer Res. 2000; 60: 1541–45.
210. Jiang BH, Jiang G, Zheng JZ, Lu Z, Hunter T, Vogt PK. Phosphatidylinositol 3-kinase signaling controls levels of hypoxia-inducible factor 1. Cell Growth Differ. 2001; 12: 363–69.
211. Koul D, Shen R, Garyali A, Ke LD, Liu TJ, Yung WK. MMAC/PTEN tumor suppressor gene egulates vascular endothelial growth factor-mediated angiogenesis in prostate cancer. Int J Oncol. 2002; 21: 469–75.

212. Gomez-Manzano C, Fueyo J, Jiang H, Glass TL, Lee HY, Hu M, Liu JL, Jasti SL, Liu TJ, Conrad CA, Yung WK. Mechanisms underlying PTEN regulation of vascular endothelial growth factor and angiogenesis. Ann Neurol. 2003; 53: 109–17.
213. Jiang YA, Fan LF, Jiang CQ, Zhang YY, Luo HS, Tang ZJ, Xia D, Wang M. Expression and significance of PTEN, hypoxia-inducible factor-1 alpha in colorectal adenoma and adenocarcinoma. World J Gastroenterol. 2003; 9: 491–94.
214. Sharpless NE, DePinho RA. The INK4A/ARF locus and its two gene products. Curr Opin Genet Dev. 1999; 9: 22–30.
215. Baylin SB, Herman JG, Graff JR, Vertino PM, Issa JP. Alterations in DNA methylation: a fundamental aspect of neoplasia. Adv Cancer Res. 1998; 72: 141–96.
216. David K, Ollert MW, Vollmert C, Heiligtag S, Eickhoff B, Erttmann R, Bredehorst R, Vogel CW. Human natural immunoglobulin M antibodies induce apoptosis of human neuroblastoma cells by binding to a Mr 260,000 antigen. Cancer Res. 1999; 59: 3768–75.
217. Miki K, Shimizu E, Yano S, Tani K, Sone S. Demethylation by 5-aza-2'-deoxycytidine (5-azadC) of p16INK4A gene results in downregulation of vascular endothelial growth factor expression in human lung cancer cell lines. Oncol Res. 2000; 12: 335–42.
218. Gibson SL, Dai CY, Lee HW, DePinho RA, Gee MS, Lee WM, Furth EE, Brensinger C, Enders GH. Inhibition of colon tumor progression and angiogenesis by the Ink4a/Arf locus. Cancer Res. 2003; 63: 742–46.
219. Senoo M, Matsumura Y, Habu S. TAp63gamma (p51A) and dNp63alpha (p73L), two major isoforms of the p63 gene, exert opposite effects on the vascular endothelial growth factor (VEGF) gene expression. Oncogene. 2002; 21: 2455–65.
220. Kawai H, Li H, Chun P, Avraham S, Avraham HK. Direct interaction between BRCA1 and the estrogen receptor regulates vascular endothelial growth factor (VEGF) transcription and secretion in breast cancer cells. Oncogene. 2002; 21: 7730–39.
221. Dent P, Yacoub A, Contessa J, Caron R, Amorino G, Valerie K, Hagan MP, Grant S, Schmidt-Ullrich R. Stress and radiation-induced activation of multiple intracellular signaling pathways. Radiat Res. 2003; 159: 283–300.
222. Danilkovitch-Miagkova A. Oncogenic signaling pathways activated by RON receptor tyrosine kinase. Curr Cancer Drug Targets. 2003 Feb; 3(1): 31–40. Review. Erratum in: Curr Cancer Drug Targets. 2003; 3: 161.
223. Ang KK, Berkey BA, Tu X, Zhang HZ, Katz R, Hammond EH, Fu KK, Milas L. Impact of epidermal growth factor receptor expression on survival and pattern of relapse in patients with advanced head and neck carcinoma. Cancer Res. 2002; 62: 7350–56.
224. Arteaga CL. Overview of epidermal growth factor receptor biology and its role as a therapeutic target in human neoplasia. Semin Oncol. 2002; 29(5 Suppl 14): 3–9.
225. Fukuda R, Hirota K, Fan F, Jung YD, Ellis LM, Semenza GL. Insulin-like growth factor 1 induces hypoxia-inducible factor 1-mediated vascular endothelial growth factor expression, which is dependent on MAP kinase and phosphatidylinositol 3-kinase signaling in colon cancer cells. J Biol Chem. 2002; 277: 38205–11.
226. Tanaka T, Kanai H, Sekiguchi K, Aihara Y, Yokoyama T, Arai M, Kanda T, Nagai R, Kurabayashi M. Induction of VEGF gene transcription by IL-1 beta is mediated through stress-activated MAP kinases and Sp1 sites in cardiac myocytes. J Mol Cell Cardiol. 2000; 32: 1955–67.
227. Vaupel P, Kallinowski F, Okunieff P. Blood flow, oxygen and nutrient supply, and metabolic microenvironment of human tumors: a review. Cancer Res. 1989; 49: 6449–65.
228. Intaglietta M, Johnson PC, Winslow RM. Microvascular and tissue oxygen distribution. Cardiovasc Res. 1996; 32: 632–43.
229. Jain RK. Determinants of tumor blood flow: a review. Cancer Res. 1988; 48: 2641–58.
230. Shweiki D, Itin A, Soffer D, Keshet E. Vascular endothelial growth factor induced by hypoxia may mediate hypoxia-initiated angiogenesis. Nature. 1992; 359(6398): 843–45.
231. Minchenko A, Bauer T, Salceda S, Caro J. Hypoxic stimulation of vascular endothelial growth factor expression in vitro and in vivo. Lab Invest. 1994; 71: 374–79.

232. Shima DT, Adamis AP, Ferrara N, Yeo KT, Yeo TK, Allende R, Folkman J, D'Amore PA. Hypoxic induction of endothelial cell growth factors in retinal cells: identification and characterization of vascular endothelial growth factor (VEGF) as the mitogen. Mol Med. 1995; 1: 182–93.
233. Levy AP, Levy NS, Wegner S, Goldberg MA. Transcriptional regulation of the rat vascular endothelial growth factor gene by hypoxia. J Biol Chem. 1995; 270: 13333–40.
234. Liu Y, Cox SR, Morita T, Kourembanas S. Hypoxia regulates vascular endothelial growth factor gene expression in endothelial cells. Identification of a 5' enhancer. Circ Res. 1995; 77: 638–43.
235. Madan A, Curtin PT. A 24-base-pair sequence 3' to the human erythropoietin gene contains a hypoxia-responsive transcriptional enhancer. Proc Natl Acad Sci USA. 1993; 90: 3928–32.
236. Cockman ME, Masson N, Mole DR, Jaakkola P, Chang GW, Clifford SC, Maher ER, Pugh CW, Ratcliffe PJ, Maxwell PH. Hypoxia inducible factor-alpha binding and ubiquitylation by the von Hippel-Lindau tumor suppressor protein. J Biol Chem. 2000; 275: 25733–41.
237. Ohh M, Park CW, Ivan M, Hoffman MA, Kim TY, Huang LE, Pavletich N, Chau V, Kaelin WG. Ubiquitination of hypoxia-inducible factor requires direct binding to the beta-domain of the von Hippel-Lindau protein. Nat Cell Biol. 2000; 2: 423–27.
238. Kamura T, Sato S, Iwai K, Czyzyk-Krzeska M, Conaway RC, Conaway JW. Activation of HIF1alpha ubiquitination by a reconstituted von Hippel-Lindau (VHL) tumor suppressor complex. Proc Natl Acad Sci USA. 2000; 97: 10430–35.
239. Maxwell PH, Wiesener MS, Chang GW, Clifford SC, Vaux EC, Cockman ME, Wykoff CC, Pugh CW, Maher ER, Ratcliffe PJ. The tumour suppressor protein VHL targets hypoxia-inducible factors for oxygen-dependent proteolysis. Nature. 1999; 399(6733): 271–75.
240. Jaakkola P, Mole DR, Tian YM, Wilson MI, Gielbert J, Gaskell SJ, von Kriegsheim A, Hebestreit HF, Mukherji M, Schofield CJ, Maxwell PH, Pugh CW, Ratcliffe PJ. Targeting of HIF-alpha to the von Hippel-Lindau ubiquitylation complex by O2-regulated prolyl hydroxylation. Science. 2001; 292(5516): 468–72
241. Forsythe JA, Jiang BH, Iyer NV, Agani F, Leung SW, Koos RD, Semenza GL. Activation of vascular endothelial growth factor gene transcription by hypoxia-inducible factor 1. Mol Cell Biol. 1996; 16: 4604–13.
242. Zhong H, De Marzo AM, Laughner E, Lim M, Hilton DA, Zagzag D, Buechler P, Isaacs WB, Semenza GL, Simons JW. Overexpression of hypoxia-inducible factor 1alpha in common human cancers and their metastases. Cancer Res. 1999; 59: 5830–35.
243. Talks KL, Turley H, Gatter KC, Maxwell PH, Pugh CW, Ratcliffe PJ, Harris AL. The expression and distribution of the hypoxia-inducible factors HIF-1alpha and HIF-2alpha in normal human tissues, cancers, and tumor-associated macrophages. Am J Pathol. 2000; 157: 411–21.
244. Ikeda E, Achen MG, Breier G, Risau W. Hypoxia-induced transcriptional activation and increased mRNA stability of vascular endothelial growth factor in C6 glioma cells. J Biol Chem. 1995; 270: 19761–66.
245. Goldberg-Cohen I, Furneauxb H, Levy AP. A 40-bp RNA element that mediates stabilization of vascular endothelial growth factor mRNA by HuR. J Biol Chem. 2002; 277: 13635–40.
246. Akiri G, Nahari D, Finkelstein Y, Le SY, Elroy-Stein O, Levi BZ. Regulation of vascular endothelial growth factor (VEGF) expression is mediated by internal initiation of translation and alternative initiation of transcription. Oncogene. 1998; 17: 227–36.
247. Miller DL, Dibbens JA, Damert A, Risau W, Vadas MA, Goodall GJ. The vascular endothelial growth factor mRNA contains an internal ribosome entry site. FEBS Lett. 1998; 434: 417–20.
248. Wike-Hooley JL, Haveman J, Reinhold HS. The relevance of tumour pH to the treatment of malignant disease. Radiother Oncol. 1984; 2: 343–66.
249. Tannock IF, Rotin D. Acid pH in tumors and its potential for therapeutic exploitation. Cancer Res. 1989; 49: 4373–84.
250. Fukumura D, Xu L, Chen Y, Gohongi T, Seed B, Jain RK. Hypoxia and acidosis independently up-regulate vascular endothelial growth factor transcription in brain tumors in vivo. Cancer Res. 2001; 61: 6020–24.

251. Xu L, Fukumura D, Jain RK. Acidic extracellular pH induces vascular endothelial growth factor (VEGF) in human glioblastoma cells via ERK1/2 MAPK signaling pathway: mechanism of low pH-induced VEGF. J Biol Chem. 2002; 277: 11368–74.
252. Brooks SE, Gu X, Kaufmann PM, Marcus DM, Caldwell RB. Modulation of VEGF production by pH and glucose in retinal Muller cells. Curr Eye Res. 1998; 17: 875–82.
253. Spector JA, Mehrara BJ, Greenwald JA, Saadeh PB, Steinbrech DS, Bouletreau PJ, Smith LP, Longaker MT. Osteoblast expression of vascular endothelial growth factor is modulated by the extracellular microenvironment. Am J Physiol Cell Physiol. 2001; 280: C72–80.
254. D'Arcangelo D, Facchiano F, Barlucchi LM, Melillo G, Illi B, Testolin L, Gaetano C, Capogrossi MC. Acidosis inhibits endothelial cell apoptosis and function and induces basic fibroblast growth factor and vascular endothelial growth factor expression. Circ Res. 2000; 86: 312–18.
255. Jang A, Hill RP. An examination of the effects of hypoxia, acidosis, and glucose starvation on the expression of metastasis-associated genes in murine tumor cells. Clin Exp Metastasis. 1997; 15: 469–83.
256. Scott PA, Gleadle JM, Bicknell R, Harris AL. Role of the hypoxia sensing system, acidity and reproductive hormones in the variability of vascular endothelial growth factor induction in human breast carcinoma cell lines. Int J Cancer. 1998; 75: 706–12.
257. Stein I, Neeman M, Shweiki D, Itin A, Keshet E. Stabilization of vascular endothelial growth factor mRNA by hypoxia and hypoglycemia and coregulation with other ischemia-induced genes. Mol Cell Biol. 1995; 15: 5363–68.
258. Damert A, Machein M, Breier G, Fujita MQ, Hanahan D, Risau W, Plate KH. Up-regulation of vascular endothelial growth factor expression in a rat glioma is conferred by two distinct hypoxia-driven mechanisms. Cancer Res. 1997; 57: 3860–64.
259. Claffey KP, Shih SC, Mullen A, Dziennis S, Cusick JL, Abrams KR, Lee SW, Detmar M. Identification of a human VPF/VEGF 3' untranslated region mediating hypoxia-induced mRNA stability. Mol Biol Cell. 1998; 9: 469–81.
260. Shi Q, Le X, Wang B, Abbruzzese JL, Xiong Q, He Y, Xie K. Regulation of vascular endothelial growth factor expression by acidosis in human cancer cells. Oncogene. 2001; 20: 3751–56.
261. Huang S, Robinson JB, Deguzman A, Bucana CD, Fidler IJ. Blockade of nuclear factor-kappaB signaling inhibits angiogenesis and tumorigenicity of human ovarian cancer cells by suppressing expression of vascular endothelial growth factor and interleukin 8. Cancer Res. 2000; 60: 5334–39.
262. Xiong HQ, Abbruzzese JL, Lin E, Wang L, Zheng L, Xie K. NF-kappaB activity blockade impairs the angiogenic potential of human pancreatic cancer cells. Int J Cancer. 2004; 108: 181–88.
263. Xie K, Fidler IJ. Therapy of cancer metastasis by activation of the inducible nitric oxide synthase. Cancer Metastasis Rev. 1998; 17: 55–75.
264. Dreher D, Junod AF. Role of oxygen free radicals in cancer development. Eur J Cancer. 1996; 32A: 30–38.
265. Lala PK, Chakraborty C. Role of nitric oxide in carcinogenesis and tumour progression. Lancet Oncol. 2001; 2: 149–56.
266. Cianchi F, Cortesini C, Fantappie O, Messerini L, Schiavone N, Vannacci A, Nistri S, Sardi I, Baroni G, Marzocca C, Perna F, Mazzanti R, Bechi P, Masini E. Inducible nitric oxide synthase expression in human colorectal cancer: correlation with tumor angiogenesis. Am J Pathol. 2003; 162: 793–801.
267. Song ZJ, Gong P, Wu YE. Relationship between the expression of iNOS, VEGF, tumor angiogenesis and gastric cancer. World J Gastroenterol. 2002; 8: 591–95.
268. Gallo O, Masini E, Morbidelli L, Franchi A, Fini-Storchi I, Vergari WA, Ziche M. Role of nitric oxide in angiogenesis and tumor progression in head and neck cancer. J Natl Cancer Inst. 1998; 90: 587–96.

269. Donnini S, Ziche M. Constitutive and inducible nitric oxide synthase: role in angiogenesis. Antioxid Redox Signal. 2002; 4: 817–23.
270. Kimura H, Esumi H. Reciprocal regulation between nitric oxide and vascular endothelial growth factor in angiogenesis. Acta Biochim Pol. 2003; 50: 49–59.
271. Uneda S, Hata H, Matsuno F, Nagasaki A, Harada N, Mitsuya Y, Matsuzaki H, Mitsuya H. A nitric oxide synthase inhibitor, N(G)-nitro-l-arginine-methyl-ester, exerts potent antiangiogenic effects on lasmacytoma in a newly established multiple myeloma severe combined immunodeficient mouse model. Br J Haematol. 2003; 120: 396–404.
272. Frank S, Madlener M, Pfeilschifter J, Werner S. Induction of inducible nitric oxide synthase and its corresponding tetrahydrobiopterin-cofactor-synthesizing enzyme GTP-cyclohydrolase I during cutaneous wound repair. J Invest Dermatol. 1998; 111: 1058–64.
273. Frank S, Stallmeyer B, Kampfer H, Kolb N, Pfeilschifter J. Nitric oxide triggers enhanced induction of vascular endothelial growth factor expression in cultured keratinocytes (HaCaT) and during cutaneous wound repair. FASEB J. 1999; 13: 2002–14.
274. Frank S, Stallmeyer B, Kampfer H, Schaffner C, Pfeilschifter J. Differential regulation of vascular endothelial growth factor and its receptor fms-like-tyrosine kinase is mediated by nitric oxide in rat renal mesangial cells. Biochem J. 1999; 338(Pt 2): 367–74.
275. Dulak J, Jozkowicz A, Dembinska-Kiec A, Guevara I, Zdzienicka A, Zmudzinska-Grochot D, Florek I, Wojtowicz A, Szuba A, Cooke JP. Nitric oxide induces the synthesis of vascular endothelial growth factor by rat vascular smooth muscle cells. Arterioscler Thromb Vasc Biol. 2000; 20: 659–66.
276. Leibovich SJ, Chen JF, Pinhal-Enfield G, Belem PC, Elson G, Rosania A, Ramanathan M, ontesinos C, Jacobson M, Schwarzschild MA, Fink JS, Cronstein B. Synergistic up-regulation of vascular endothelial growth factor expression in murine macrophages by adenosine A(2A) receptor agonists and endotoxin. Am J Pathol. 2002; 160: 2231–44.
277. Konopka TE, Barker JE, Bamford TL, Guida E, Anderson RL, Stewart AG. Nitric oxide synthase II gene disruption: implications for tumor growth and vascular endothelial growth factor production. Cancer Res. 2001; 61: 3182–87.
278. Chin K, Kurashima Y, Ogura T, Tajiri H, Yoshida S, Esumi H. Induction of vascular endothelial growth factor by nitric oxide in human glioblastoma and hepatocellular carcinoma cells. Oncogene. 1997; 15: 437–42.
279. Kimura H, Ogura T, Kurashima Y, Weisz A, Esumi H. Effects of nitric oxide donors on vascular endothelial growth factor gene induction. Biochem Biophys Res Commun. 2002; 296: 976–82.
280. Sandau KB, Faus HG, Brune B. Induction of hypoxia-inducible-factor 1 by nitric oxide is mediated via the PI3K pathway. Biochem Biophys Res Commun. 2000; 278: 263–67.
281. Sandau KB, Fandrey J, Brune B. Accumulation of HIF-1alpha under the influence of nitric oxide. Blood. 2001; 97: 1009–15.
282. Lander HM, Milbank AJ, Tauras JM, Hajjar DP, Hempstead BL, Schwartz GD, Kraemer RT, Mirza UA, Chait BT, Burk SC, Quilliam LA. Redox regulation of cell signalling. Nature 1996; 381: 380–81.
283. Mazure NM, Brahimi-Horn MC, Pouyssegur J. Protein kinases and the hypoxia-inducible factor-1, two switches in angiogenesis. Curr Pharm Des. 2003; 9: 531–41.
284. Chen Y, Maguire T, Hileman RE, Fromm JR, Esko JD, Linhardt RJ, Marks RM. Dengue virus infectivity depends on envelope protein binding to target cell heparan sulfate. Nat Med. 1997; 3: 866–71.
285. Tuder RM, Flook BE, Voelkel NF. Increased gene expression for VEGF and the VEGF receptors KDR/Flk and Flt in lungs exposed to acute or to chronic hypoxia: modulation of gene expression by nitric oxide. J Clin Invest. 1995; 95: 1798–807.
286. Liu Y, Christou H, Morita T, Laughner E, Semenza GL, Kourembanas S. Carbon monoxide and nitric oxide suppress the hypoxic induction of vascular endothelial growth factor gene via the 5' enhancer. J Biol Chem. 1998; 273: 15257–62.

287. Yin JH, Yang DI, Ku G, Hsu CY. iNOS expression inhibits hypoxia-inducible factor-1 activity. Biochem Biophys Res Commun. 2000; 279: 30–34.
288. Ghiso N, Rohan RM, Amano S, Garland R, Adamis AP. Suppression of hypoxia-associated vascular endothelial growth factor gene expression by nitric oxide via cGMP. Invest Ophthalmol Vis Sci. 1999; 40: 1033–39.
289. Xie K, Huang S. Contribution of nitric oxide-mediated apoptosis to cancer metastasis inefficiency. Free Radic Biol Med. 2003; 34: 969–86.
290. Ambs S, Merriam WG, Ogunfusika MO, Bennett WP, Ishibe N, Hussain SP, Tzeng EE, Geller DA, Billiar TR, Harris CC. p53 and vascular endothelial growth factor regulate tumor growth of NOS2-expressing human carcinoma cells. Nat Med. 1998; 4: 1371–76.
291. Tsurumi Y, Murohara T, Krasinski K, Chen D, Witzenbichler B, Kearney M, Couffinhal T, Isner JM. Reciprocal relation between VEGF and NO in the regulation of endothelial integrity. Nat Med. 1997; 3: 879–86.
292. Finkel T, Holbrook NJ. Oxidants, oxidative stress and the biology of ageing. Nature. 2000; 408(6809): 239–47.
293. Sundaresan M, Yu ZX, Ferrans VJ, Sulciner DJ, Gutkind JS, Irani K, Goldschmidt-Clermont PJ, Finkel T. Regulation of reactive-oxygen-species generation in fibroblasts by Rac1. Biochem J. 1996; 318(Pt 2): 379–82.
294. Woo CH, Eom YW, Yoo MH, You HJ, Han HJ, Song WK, Yoo YJ, Chun JS, Kim JH. Tumor necrosis factor-alpha generates reactive oxygen species via a cytosolic phospholipase A2-linked cascade. J Biol Chem. 2000; 275: 32357–62.
295. Sundaresan M, Yu ZX, Ferrans VJ, Irani K, Finkel T. Requirement for generation of H_2O_2 for platelet-derived growth factor signal transduction. Science. 1995; 270: 296–99.
296. Pani G, Colavitti R, Bedogni B, Anzevino R, Borrello S, Galeotti T. A redox signaling mechanism for density-dependent inhibition of cell growth. J Biol Chem. 2000; 275: 38891–99.
297. Pani G, Colavitti R, Borrello S, Galeotti T. Endogenous oxygen radicals modulate protein tyrosine phosphorylation and JNK-1 activation in lectin-stimulated thymocytes. Biochem J. 2000; 347(Pt 1): 173–81.
298. Finkel T. Oxygen radicals and signaling. Curr Opin Cell Biol. 1998; 10: 248–53.
299. Rhee SG, Bae YS, Lee SR, Kwon J. Hydrogen peroxide: a key messenger that modulates protein phosphorylation through cysteine oxidation. Sci STKE. 2000; 2000(53): PE1.
300. Maulik N, Das DK. Redox signaling in vascular angiogenesis. Free Radic Biol Med. 2002; 33: 1047–60.
301. Deshpande SS, Irani K. Oxidant signalling in carcinogenesis: a commentary. Hum Exp Toxicol. 2002; 21: 63–64.
302. Arbiser JL, Petros J, Klafter R, Govindajaran B, McLaughlin ER, Brown LF, Cohen C, Moses M, Kilroy S, Arnold RS, Lambeth JD. Reactive oxygen generated by Nox1 triggers the angiogenic switch. Proc Natl Acad Sci USA. 2002; 99: 715–20.
303. Zhu JW, Yu BM, Ji YB, Zheng MH, Li DH. Upregulation of vascular endothelial growth factor by hydrogen peroxide in human colon cancer. World J Gastroenterol. 2002; 8: 153–57.
304. Chua CC, Hamdy RC, Chua BH. Upregulation of vascular endothelial growth factor by H_2O_2 in rat heart endothelial cells. Free Radic Biol Med. 1998; 25: 891–97.
305. Gabbita SP, Robinson KA, Stewart CA, Floyd RA, Hensley K. Redox regulatory mechanisms of cellular signal transduction. Arch Biochem Biophys. 2000; 376: 1–13.
306. Michiels C, Minet E, Mottet D, Raes M. Regulation of gene expression by oxygen: NF-kappaB and HIF-1, two extremes. Free Radic Biol Med. 2002; 33: 1231–42.
307. Maulik N. Redox signaling of angiogenesis. Antioxid Redox Signal. 2002; 4: 805–15.
308. Sen CK, Khanna S, Babior BM, Hunt TK, Ellison EC, Roy S. Oxidant-induced vascular endothelial growth factor expression in human keratinocytes and cutaneous wound healing. J Biol Chem. 2002; 277: 33284–90.
309. Gao N, Ding M, Zheng JZ, Zhang Z, Leonard SS, Liu KJ, Shi X, Jiang BH. Vanadate-induced expression of hypoxia-inducible factor 1 alpha and vascular endothelial growth factor through phosphatidylinositol 3-kinase/Akt pathway and reactive oxygen species. J Biol Chem. 2002; 277: 31963–71.

310. Esposito F, Chirico G, Montesano Gesualdi N, Posadas I, Ammendola R, Russo T, Cirino G, Cimino F. Protein kinase B activation by reactive oxygen species is independent of tyrosine kinase receptor phosphorylation and requires SRC activity. J Biol Chem. 2003; 278: 20828–34.
311. Kosmidou I, Xagorari A, Roussos C, Papapetropoulos A. Reactive oxygen species stimulate VEGF production from C(2)C(12) skeletal myotubes through a PI3K/Akt pathway. Am J Physiol Lung Cell Mol Physiol. 2001; 280: L585–92.
312. Colavitti R, Pani G, Bedogni B, Anzevino R, Borrello S, Waltenberger J, Galeotti T. Reactive oxygen species as downstream mediators of angiogenic signaling by vascular endothelial growth factor receptor-2/KDR.J Biol Chem. 2002; 277: 3101–08.
313. Frank S, Hubner G, Breier G, Longaker MT, Greenhalgh DG, Werner S. Regulation of vascular endothelial growth factor expression in cultured keratinocytes: implications for normal and impaired wound healing. J Biol Chem. 1995; 270: 12607–13.
314. Pertovaara L, Kaipainen A, Mustonen T, Orpana A, Ferrara N, Saksela O, Alitalo K. Vascular endothelial growth factor is induced in response to transforming growth factor-beta in fibroblastic and epithelial cells. J Biol Chem. 1994; 269: 6271–74.
315. Stavri GT, Zachary IC, Baskerville PA, Martin JF, Erusalimsky JD. Basic fibroblast growth factor upregulates the expression of vascular endothelial growth factor in vascular smooth muscle cells: synergistic interaction with hypoxia. Circulation. 1995; 92: 11–14.
316. Deroanne CF, Hajitou A, Calberg-Bacq CM, Nusgens BV, Lapiere CM. Angiogenesis by fibroblast growth factor 4 is mediated through an autocrine up-regulation of vascular endothelial growth factor expression. Cancer Res. 1997; 57: 5590–97.
317. Wang D, Huang HJ, Kazlauskas A, Cavenee WK. Induction of vascular endothelial growth factor expression in endothelial cells by platelet-derived growth factor through the activation of phosphatidylinositol 3-kinase. Cancer Res. 1999; 59: 1464–72.
318. Wojta J, Kaun C, Breuss JM, Koshelnick Y, Beckmann R, Hattey E, Mildner M, Weninger W, Nakamura T, Tschachler E, Binder BR. Hepatocyte growth factor increases expression of vascular endothelial growth factor and plasminogen activator inhibitor-1 in human keratinocytes and the vascular endothelial growth factor receptor flk-1 in human endothelial cells. Lab Invest. 1999; 79: 427–38.
319. Li J, Perrella MA, Tsai JC, Yet SF, Hsieh CM, Yoshizumi M, Patterson C, Endege WO, Zhou F, Lee ME. Induction of vascular endothelial growth factor gene expression by interleukin-1 beta in rat aortic smooth muscle cells. J Biol Chem. 1995; 270: 308–12.
320. Ben-Av P, Crofford LJ, Wilder RL, Hla T. Induction of vascular endothelial growth factor expression in synovial fibroblasts by prostaglandin E and interleukin-1: a potential mechanism for inflammatory angiogenesis. FEBS Lett. 1995; 372: 83–87.
321. Cohen T, Nahari D, Cerem LW, Neufeld G, Levi BZ. Interleukin 6 induces the expression of vascular endothelial growth factor. J Biol Chem. 1996; 271: 736–41.
322. Warren RS, Yuan H, Matli MR, Ferrara N, Donner DB. Induction of vascular endothelial growth factor by insulin-like growth factor 1 in colorectal carcinoma. J Biol Chem. 1996; 271: 29483–88.
323. Samaniego F, Markham PD, Gendelman R, Watanabe Y, Kao V, Kowalski K, Sonnabend JA, Pintus A, Gallo RC, Ensoli B. Vascular endothelial growth factor and basic fibroblast growth factor present in Kaposi's sarcoma (KS) are induced by inflammatory cytokines and synergize to promote vascular permeability and KS lesion development. Am J Pathol. 1998; 152: 1433–43.
324. Yamaguchi R, Yano H, Iemura A, Ogasawara S, Haramaki M, Kojiro M. Expression of vascular endothelial growth factor in human hepatocellular carcinoma. Hepatology. 1998; 28: 68–77.
325. Bolling B, Fandrey J, Frosch PJ, Acker H. VEGF production, cell proliferation and apoptosis of human IGR 1 melanoma cells under nIFN-alpha/beta and rIFN-gamma treatment. Exp Dermatol. 2000; 9: 327–35.
326. Xiong M, Elson G, Legarda D, Leibovich SJ. Production of vascular endothelial growth factor by murine macrophages: regulation by hypoxia, lactate, and the inducible nitric oxide synthase pathway. Am J Pathol. 1998; 153: 587–98.

327. Yamamoto S, Yasui W, Kitadai Y, Yokozaki H, Haruma K, Kajiyama G, Tahara E. Expression of vascular endothelial growth factor in human gastric carcinomas. Pathol Int. 1998; 48: 499–506.
328. Salven P, Anttonen K, Repo H, Joensuu H, Orpana A. Endotoxins induce and interferon alpha suppresses vascular endothelial growth factor (VEGF) production in human peripheral blood mononuclear cells. FASEB J. 2001; 15: 1318–20.
329. von Marschall Z, Scholz A, Cramer T, Schafer G, Schirner M, Oberg K, Wiedenmann B, Hocker M, Rosewicz S. Effects of interferon alpha on vascular endothelial growth factor gene transcription and tumor angiogenesis. J Natl Cancer Inst. 2003; 95: 437–48.
330. Tosato G. Interferon-alpha is implicated in the transcriptional regulation of vascular endothelial growth factor. J Natl Cancer Inst. 2003; 95: 420–21.
331. Mori K, Tani M, Kamata K, Kawamura H, Urata Y, Goto S, Kuwano M, Shibata S, Kondo T. Mitogen-activated protein kinase, ERK1/2, is essential for the induction of vascular endothelial growth factor by ionizing radiation mediated by activator protein-1 in human glioblastoma cells. Free Radic Res. 2000; 33: 157–66.
332. Jung YD, Nakano K, Liu W, Gallick GE, Ellis LM. Extracellular signal-regulated kinase activation is required for up-regulation of vascular endothelial growth factor by serum starvation in human colon carcinoma cells. Cancer Res. 1999; 59: 4804–07.
333. Harada S, Nagy JA, Sullivan KA, Thomas KA, Endo N, Rodan GA, Rodan SB. Induction of vascular endothelial growth factor expression by prostaglandin E2 and E1 in osteoblasts. J Clin Invest. 1994; 93: 2490–96.
334. Bamba H, Ota S, Kato A, Kawamoto C, Fujiwara K. Prostaglandins up-regulate vascular endothelial growth factor production through distinct pathways in differentiated U937 cells. Biochem Biophys Res Commun. 2000; 273: 485–91.
335. Liu XH, Kirschenbaum A, Lu M, Yao S, Dosoretz A, Holland JF, Levine AC. Prostaglandin E2 induces hypoxia-inducible factor-1alpha stabilization and nuclear localization in a human prostate cancer cell line. J Biol Chem. 2002; 277: 50081–86.
336. Wakiya K, Shibuya M. Okadaic acid stimulates the expression of vascular endothelial growth factor gene. Biochem Biophys Res Commun. 1999; 265: 584–88.
337. Shweiki D, Neeman M, Itin A, Keshet E. Induction of vascular endothelial growth factor expression by hypoxia and by glucose deficiency in multicell spheroids: implications for tumor angiogenesis. Proc Natl Acad Sci USA. 1995; 92: 768–72.
338. Satake S, Kuzuya M, Miura H, Asai T, Ramos MA, Muraguchi M, Ohmoto Y, Iguchi A. Up-regulation of vascular endothelial growth factor in response to glucose deprivation. Biol Cell. 1998; 90: 161–68.
339. Beerepoot LV, Shima DT, Kuroki M, Yeo KT, Voest EE. Up-regulation of vascular endothelial growth factor production by iron chelators. Cancer Res. 1996; 56: 3747–51.
340. Clauss M, Schaper W. Vascular endothelial growth factor: A Jack-of-all-trades or a nonspecific stress gene? Circ Res. 2000; 86: 251–52.
341. Spyridopoulos I, Brogi E, Kearney M, Sullivan AB, Cetrulo C, Isner JM, Losordo DW. Vascular endothelial growth factor inhibits endothelial cell apoptosis induced by tumor necrosis factor-alpha: balance between growth and death signals. J Mol Cell Cardiol. 1997; 29: 1321–30.
342. Watanabe Y, Dvorak HF. Vascular permeability factor/vascular endothelial growth factor inhibits anchorage-disruption-induced apoptosis in microvessel endothelial cells by inducing scaffold formation. Exp Cell Res. 1997; 233: 340–49.
343. Gerber HP, McMurtrey A, Kowalski J, Yan M, Keyt BA, Dixit V, Ferrara N. Vascular endothelial growth factor regulates endothelial cell survival through the phosphatidylinositol 3'-kinase/Akt signal transduction pathway: requirement for Flk-1/KDR activation. J Biol Chem. 1998; 273: 30336–43.

Chapter 7
Incipient Events in Human Carcinogenesis: A Concept of Forerunner Genes

Bogdan Czerniak

Abstract Many common epithelial human cancers start as *in situ* expansions of cells, which exhibit almost normal phenotype. Such expansions form large plaques involving the affected mucosal membrane and are antecedent to the development of dysplasia or carcinoma *in situ*. Here we describe a whole-organ genomic mapping strategy to identify specific chromosomal regions involved in the development of early intraurothelial lesions in human bladder carcinogenesis. High resolution mapping of one such regions containing the model tumor suppressor RB1 provided evidence supporting the existence of a new class of genes termed forerunner (FR) genes. These genes map near tumor suppressors and provide a critical driving force for the early clonal expansion of neoplastic cells. The FR genes are related to tumor suppressors as they contribute to tumor development by their loss of function but their inactivation occurs prior to that of tumor suppressors such as RB1 in tumorigenesis.

Keywords Bladder cancer · Carcinogenesis · Whole-organ genetic mapping · Forerunner genes · Hyperplasia · In situ neoplasia · Loss of heterozygosity (LOH) · Polymorphism · Genetic instability · Homologous recombination

Introduction

During the last three decades, the central stage of cancer research has been taken by the studies focused on the effects of transforming and suppressor genes that gradually change normal cells into cancer cells [1]. The prototypic multi-step model of cancer development was originally constructed for an adenoma colonic carcinoma sequence [2] and similar models were subsequently proposed for virtually all common human malignancies [3–10].

B. Czerniak
Department of Pathology, Nathan W. Lassiter Distinguished Chair in Urology, Department of Pathology, The University of Texas, MD Anderson Cancer Center, Houston, TX 77030 USA
e-mail: bczernia@mdanderson.org

M. Bar-Eli (ed.), *Regulation of Gene Expression in the Tumor Environment*,

The original dual-track concept of urinary bladder carcinogenesis, postulated approximately three decades ago, was developed on the basis of clinicopathologic observations and whole-organ microscopic mapping studies of cystectomy specimens [20–22]. These early studies postulated that urothelial neoplasia progressed from precursor lesions such as hyperplasia with low to moderate dysplasia (low-grade intraurothelial neoplasia, LGIN) to severe dysplasia and carcinoma *in situ* (high-grade intraurothelial neoplasia, HGIN) and finally to invasive cancer. Furthermore, virtually every clinically evident bladder tumor was found to be associated with wide microscopically recognizable changes in the urinary bladder mucosa representing either LGIN or HGIN. It is now generally accepted that invasive bladder cancer develops by the LGIN-HGIN sequence via complex stepwise molecular events.

Fig. 7.2 (continued) **Assembly of whole-organ histologic and genetic maps.** (**a**) A gross photograph of an open cystectomy specimen showing an invasive carcinoma (*upper panel*). White arrows indicate the tumor. The mucosa was divided into 1×2 cm rectangular samples and evaluated microscopically on frozen sections stained with hemotoxylin and eosin. The results of the microscopic evaluation were recorded as a histologic map (*bottom panel*). The histologic map code is as follows: NU, normal urothelium; MD, mild dysplasia; MdD, moderate dysplasia; SD, severe dysplasia; CIS, carcinoma *in situ*; and TCC, transitional cell carcinoma. The areas of bladder mucosa that were involved by clonal allelic losses of markers D3S1541 and D12S397 are delineated by continuous and interrupted red lines, respectively. The positions of these markers on the sex-averaged recombination-based map of chromosome 3 and 12 as well as their band positions are shown on the left. The marker D12S397 shows a plaque-like clonal LOH that involved almost the entire bladder mucosa. In contrast, the marker D3S1541 involved a smaller area restricted to HGIN and invasive TCC only. (**b**) Representative microscopic samples of NU, precursor *in situ* conditions (LGIN, HGIN), and TCC are shown. For the purpose of statistical analysis intraurothelial precursor conditions were classified into two groups: low-grade intraurothelial neoplasia (mild to moderate dysplasia, LGIN) and high-grade intraurothelial neoplasia (severe dysplasia and carcinoma *in situ*, HGIN). Note the increasing nuclear atypia and architectural disorder when intraurothelial precursor conditions advanced from LGIN to HGIN and invasive stromal growth in TCC. Solid black bars within the photomicrographs indicate 50 μm. (**c**) Examples of allelic patterns for the two markers (D12S397 and D3S1541) which were tested on multiple mucosal samples (numbered 1–13) from the same cystectomy specimen resolved on polyacrylamide (PAM) and by capillary electrophoresis (CE) are illustrated. Sample #1 shows the allelic pattern of the same marker from peripheral blood lymphocytes (PBDNA) of the same patient. In capillary electrophoresis AA designates the upper allele and LA designates the lower allele. The ratios of signal intensities (RSI) were calculated for the two alleles in the tested samples (*upper row*) and also for the same allele in the tested samples compared to paired non-tumor DNA (*lower row*). RSI $\leq$0.5 for both calculations was considered indicative of LOH. Marker D12S397 showed clonal LOH in multiple samples corresponding to TCC, HGIN, LGIN, and also involved some areas of bladder mucosa with microscopically normal urothelium. Clonal allelic loss of D3S1541 was restricted to invasive TCC and adjacent HGIN. Such patterns of alterations implicate a loss of D12S397 in the early *in situ* expansion of a neoplasia clone that involved large areas of bladder mucosa. In contrast, the loss of D3S1541 occurred later than D12S397 and was associated with expansion of a successive clone having features of severe dysplasia/carcinoma *in situ* that progressed to invasive carcinoma. (Modified and reprinted with permission from S. Lee et al. *PNAS* 104(34):13732–13737, 2007.)

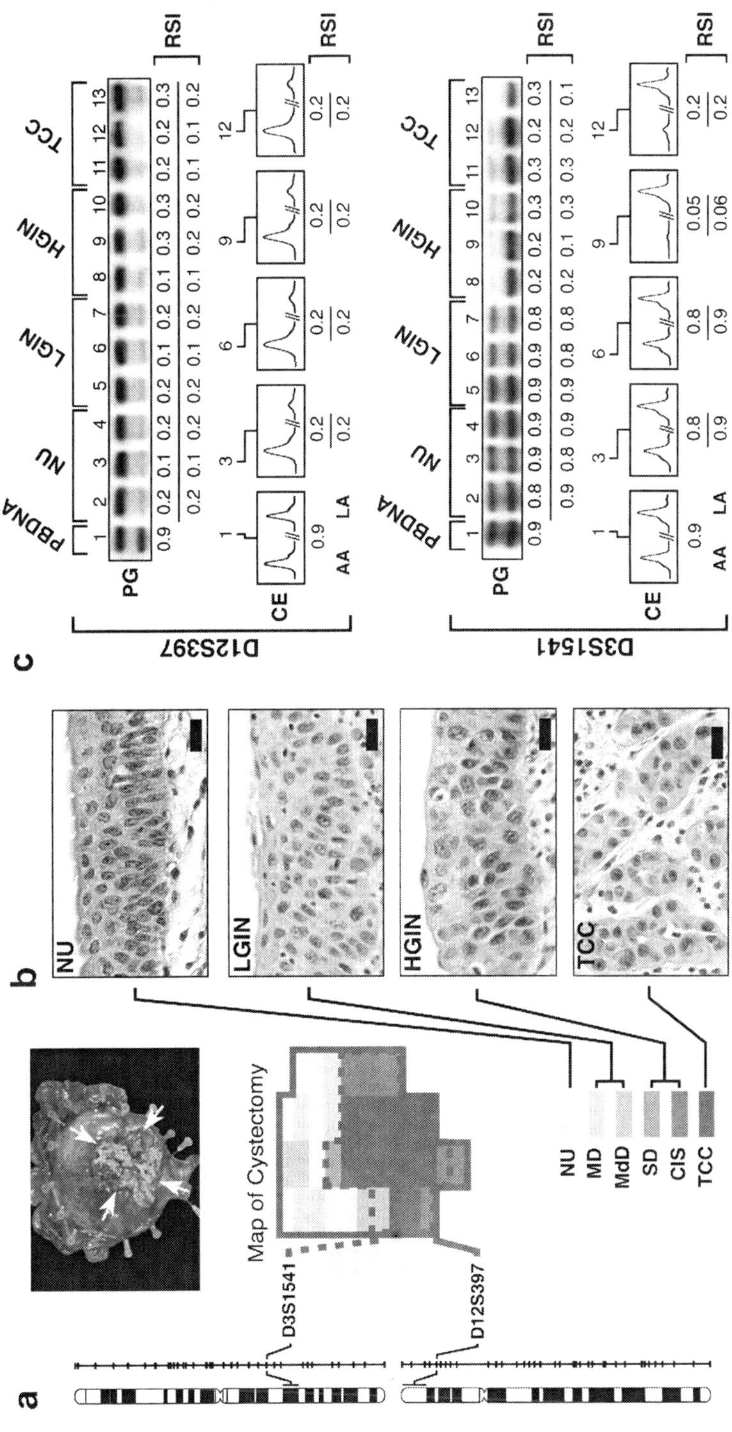

Fig. 7.2 (continued)

Whole-Organ Histologic and Genetic Mapping

Analysis of genomic imbalances can guide us to those chromosomal regions that contain genes and non-coding sequences playing a role in tumor development. In familial disorders including cancer predisposing syndromes, a cosegregation of genetic hits with diseased phenotype identifies a predisposing locus and may guide subsequent identification of a positional target gene [23, 24]. In sporadic epithelial cancers that develop from microscopically recognizable pre-neoplastic *in situ* conditions however, the early events can instead be deduced from the geographic relationship between genomic imbalance and precursor *in situ* conditions (Fig. 7.2a–c) [15, 25–31]. The similarity of alterations such as loss of the same allele or the presence of identical molecular alterations in multiple samples corresponding to precursor conditions and invasive cancer disclose their clonal relationship. Together, these identify genetic hits that provide growth advantage related to a particular clone of cells collectively referred to as clonal expansion.

We developed a strategy to identify genomic imbalances across the entire mucosa of an affected organ associated with cancer development, from *in situ* precursor conditions to invasive disease, on a total genomic scale (Fig. 7.3). We used WOHGM to identify clonal genomic imbalances such as loss of heterozygosity (LOH) or loss of polymorphism (LOP) associated with growth advantage, thus tracking the development of human bladder cancer from intraurothelial precursor lesions.

For WOHGM, ressected bladders with TCC were divided into approximately 30–60 mucosal samples, each covering 2 cm^2 of mucosal area and corresponding to microscopically normal urothelium, precursor intraurothelial conditions defined as LGIN or HGIN, or TCC. The urothelial lining was stripped by mechanical scraping, providing urothelial cell suspensions, which typically yielded 5–10 μg of high quality genomic DNA for mapping studies.

By superimposing the distribution patterns of genomic imbalances over the histologic maps of the entire mucosa the two basic patterns can be identified: scattered, in which several isolated foci are present and plaque-like where a large contiguous area of identical genetic changes are found. The scattered isolated foci showing limited clonal expansion of preneoplastic cells are typically not considered to be functionally relevant for disease development and progression (data not shown). The changes which form plaque-like areas associated with clonal *in situ* expansion of preneoplastic cells involving large areas of mucosa that encompass not only invasive cancer and precursor conditions but also adjacent areas of microscopically normal epithelium represent early events associated with the development of incipient occult phases of neoplasia. On the opposite side of the spectrum are hits restricted to invasive carcinoma and adjacent areas of severe dysplasia or carcinoma *in situ* representing late events associated with the progression to invasive cancer (Fig. 7.3).

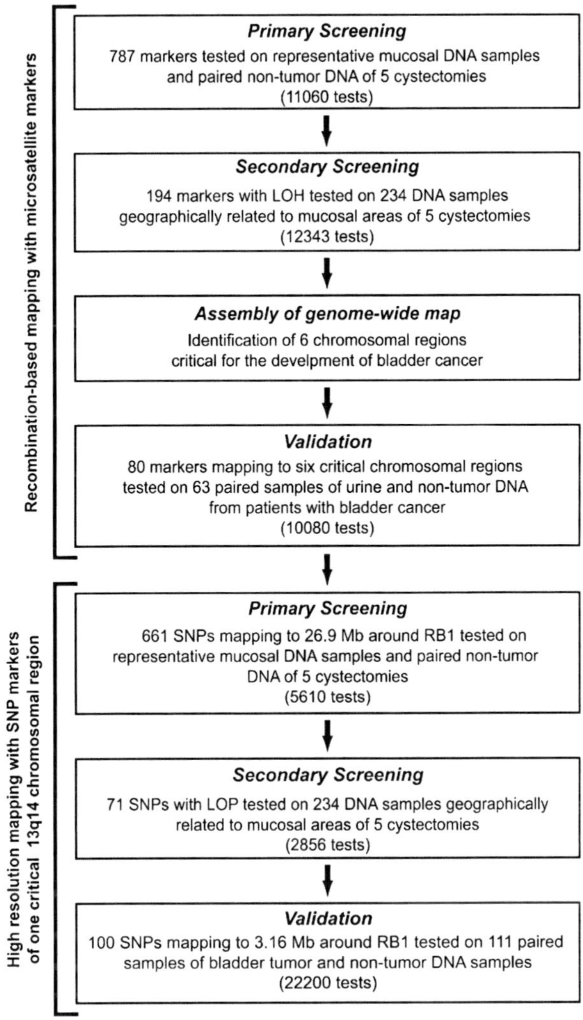

Fig. 7.3 Strategy used to construct genomic model of bladder cancer. The primary screening with hypervariable DNA markers was performed on paired samples of non-tumor and invasive tumor DNA. Markers showing LOH were selected for secondary screening on all mucosal samples of the same cystectomy. Markers mapping to autosomes 1-22 were tested on five cystectomy specimens. The pattern of LOH on chromosomes 1-22 was used to construct a genome-wide map of bladder cancer development and to identify six chromosomal regions critical for clonal expansion of *in situ* neoplasia. Finally, the high-resolution mapping was performed on one of the critical chromosomal regions containing a model tumor suppressor, *RB1*. These studies defined a minimal deleted region associated with clonal expansion of intraurothelial neoplasia around *RB1* and permitted the identification of novel target FR genes providing growth advantage for this expansion. (Reprinted with permission from T. Majewski et al. *Lab Invest* 2008.)

Genomic Model of Bladder Cancer

We used a WOHGM strategy to construct a genome-wide map of bladder cancer development from preneoplastic *in situ* conditions to invasive disease (Fig. 7.4) [16]. In order to identify those changes that were most likely critical for tumor development, we searched our data for overlapping plaques of clonal allelic losses demonstrating a geographic relationship to the two major phases of urothelial neoplasia. (Fig. 7.5a,b). The first group consisted of alterations associated with expansion of a dominant clone with no or minimal phenotypic change that involved large areas of bladder mucosa. The second group consisted of alterations associated with the development of successive clones showing a fully transformed phenotype i.e. those that were related to onset of HGIN progressing to invasive cancer. When patterns of chromosomal losses from several resected bladders were analyzed it became evident that six chromosomal regions mapping to 3q22-q24, 5q22-q31, 9q21-q22, 10q26, 13q14, and 17p13 containing well-known tumor suppressor genes were involved in more than one case and may represent six critical hits driving the development of human bladder cancer (Fig. 7.6a–d). LOH affecting at least one of the critical chromosomal regions could be identified in 98% of bladder cancers. The frequency of LOH in each of the chromosomal regions was similar in patients with clinically evident tumor and in patients with history of bladder tumor removed by a transurethral resection. The frequency of LOH was also similar in low-grade superficial and high-grade invasive tumors.

High Resolution Mapping of 13q14 Region

To identify genomic sequences predisposing to the development of genomic imbalances associated with early clonal expansion of intraurothelial neoplasia as well as to identify novel genes that may provide growth advantage for such expansion, we performed a high resolution WOHGM of one of the critical chromosomal regions mapping to 13q14 and containing the model tumor suppressor *RB1* (Fig. 7.7a–d) [30].

The pattern of losses identified by high resolution WOHGM studies defined a minimal region of 1.34 Mb associated with clonal expansion of *in situ* neoplasia around *RB1*. A loss of DNA that affected *RB1* and its flanking region was associated with clonal expansion that formed a plaque involving large areas of bladder mucosa

Fig. 7.4 (continued) of open or *solid circles* on appropriate *concentric circles* relate the alterations to a given phase of neoplasia. Only markers with LOH are positioned on the chromosomal vectors. Solid bars on outer brackets represent clusters of markers with significant LOH and denote location of putative chromosomal regions involved in urothelial neoplasia. The distances of markers on chromosomal vectors and the solid bars depicting minimal deleted regions were adjusted to fit the oval and are not drawn to scale. (Reprinted with permission from T. Majewski et al. *Lab Invest* 2008.)

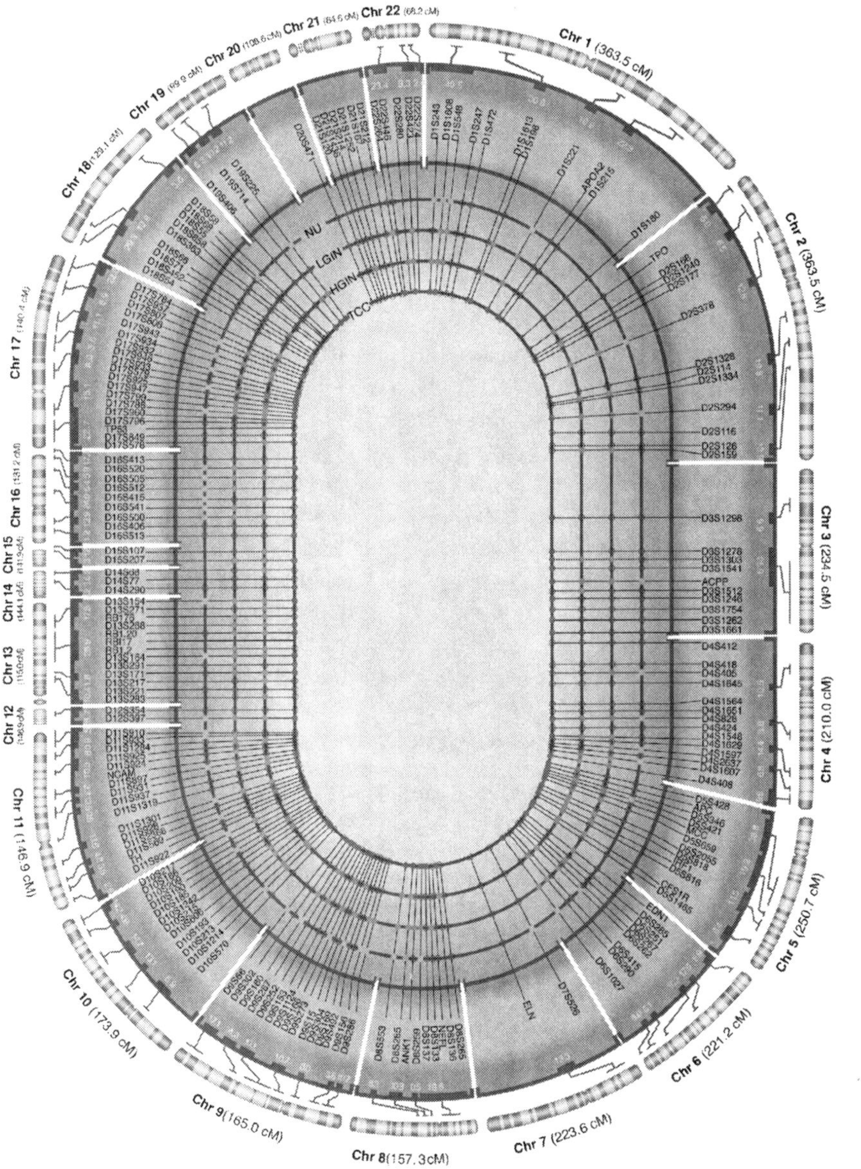

Fig. 7.4 Genome-wide map of bladder cancer progression from intraurothelial precursor conditions to invasive disease. The map was assembled on the basis of whole-organ histologic and genetic mapping of chromosomes 1-22. The outer circle represents chromosomal vectors aligned clockwise from p to q arms with positions of altered markers exhibiting LOH. The innermost concentric circles represent major phases of development and progression of urothelial neoplasia from normal urothelium (NU) through low-grade intraurothelial neoplasia (LGIN) and high-grade intraurothelial neoplasia (HGIN) to transitional cell carcinoma (TCC). *Solid circles* (•) denote statistically significant LOH of the markers defined by the LOD score analysis. *Open circles* (○) identify LOH without statistically significant association to a given stage of neoplasia. The position

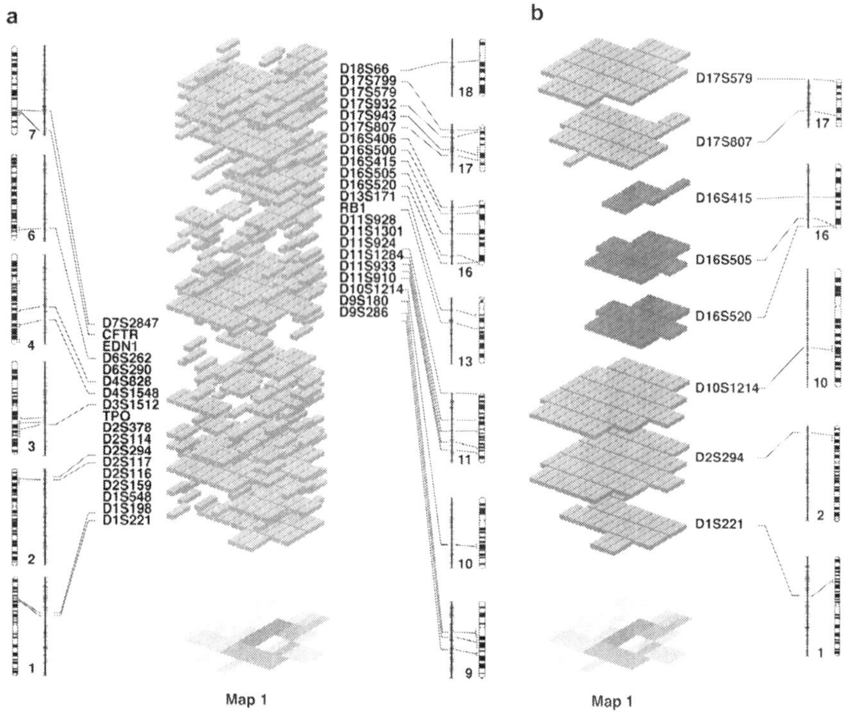

Fig. 7.5 Genome-wide pattern of LOH identified by WOHGM in a single cystectomy. (**a**) Three-dimensional display of the LOH distribution patterns in a single cystectomy specimen. The vertical axis represents sex-averaged recombination-based chromosomal maps with positions of hypervariable markers and their chromosomal location. The shaded blocks represent areas of bladder mucosa with LOH as they relate to the development of bladder cancer from *in situ* neoplasia, represented by a histologic map of the cystectomy shown at the bottom. The histologic map code is the same as in Fig. 7.2. (**b**) Clonal losses associated with expansion of *in situ* neoplasia. Chromosomal regions exhibiting allelic losses associated with early and late phases of bladder neoplasia were identified as described in Fig. 7.2. (Modified and reprinted with permission from T. Majewski et al. *Lab Invest* 2008.)

Fig. 7.6 (continued) restricted to HGIN and TCC. *Black arrows* indicate six regions critical for the development of bladder cancer. (**b**) Frequency of LOH in six critical chromosomal regions in patients with clinically evident tumor and patients with history of bladder cancer and no evidence of disease at the time of testing. (**c**) Frequency of LOH in six critical chromosomal regions in low (grade 1–2) and high (grade 3) grade TCCs. (**d**) Frequency of synchronous involvement of one or more critical chromosomal regions identified in voided urine in all 63 patients with bladder cancer. (Modified and reprinted with permission from S. Lee et al. *PNAS* 104(34):13732–13737, 2007.)

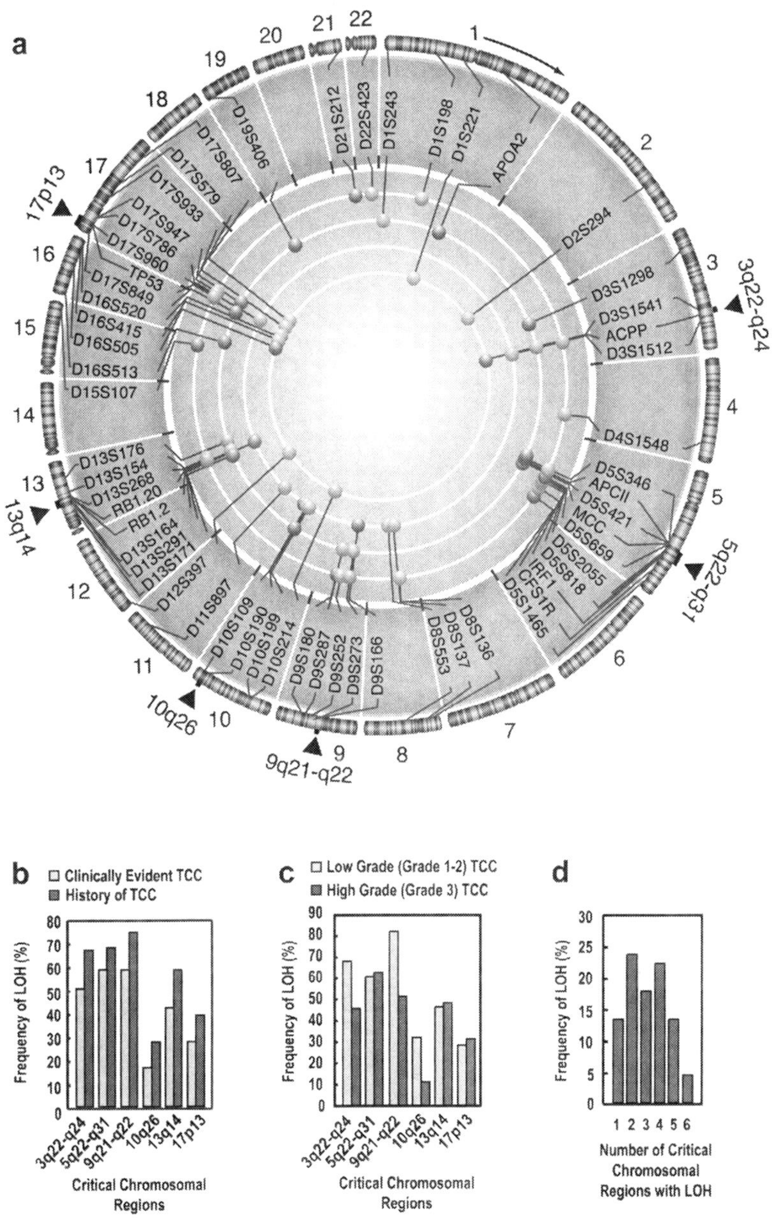

Fig. 7.6 Critical chromosomal regions involved in the development of bladder cancer. (**a**) The *outer circle* depicts the recombination-based map of chromosomes arranged clockwise from p to q arms. The four *innermost circles* represent maps of the informative cystectomies. Green dots designate markers with clonal LOH forming plaques involving microscopically normal appearing urothelium (NU), low grade intraurothelial neoplasia (LGIN), high grade intraurothelial neoplasia (HGIN), and transitional cell carcinoma (TCC). Red dots designate markers with clonal LOH

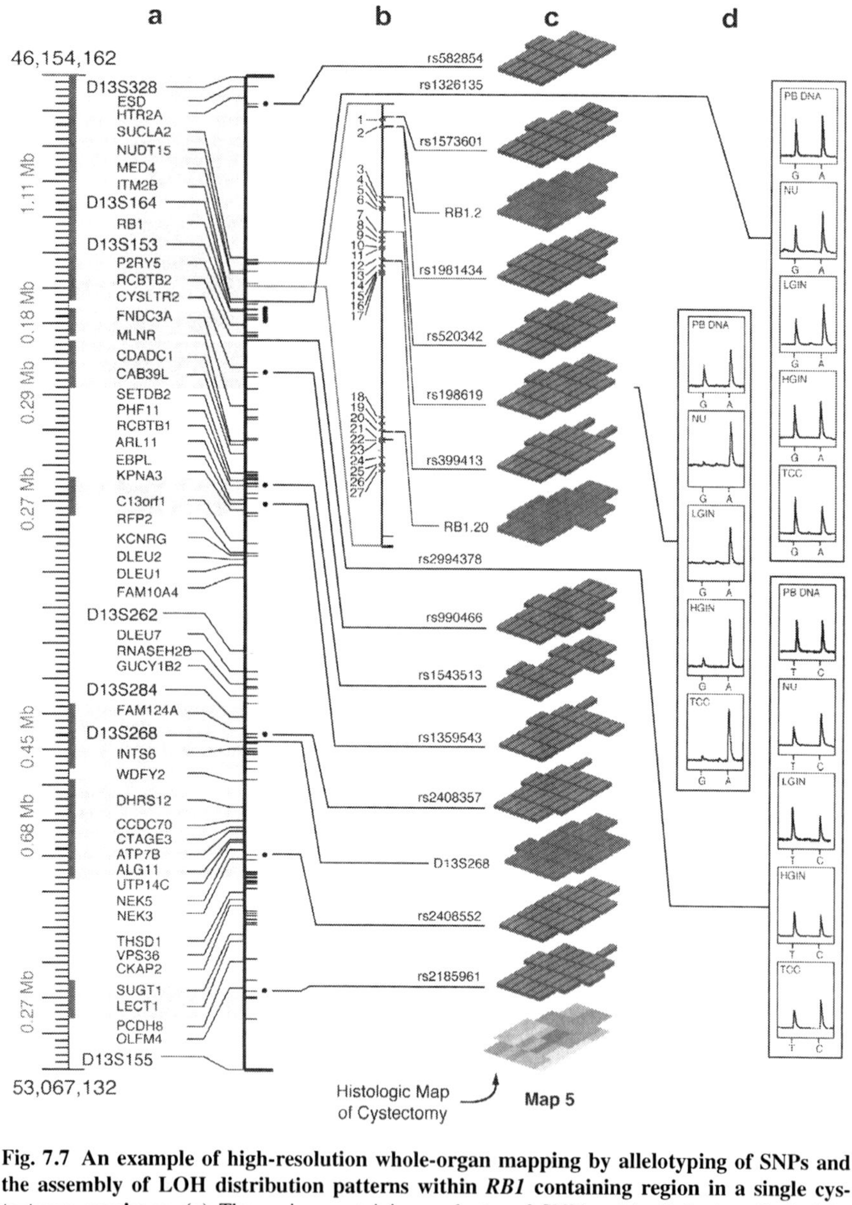

Fig. 7.7 An example of high-resolution whole-organ mapping by allelotyping of SNPs and the assembly of LOH distribution patterns within *RB1* containing region in a single cystectomy specimen. (**a**) The region containing a cluster of SNP's with allelic loss flanked by markers D13S328 and D13S155 is shown. The *bars* on the *left* side indicate the positions of all known and computationally predicted genes. The *bars* on the *right* side designate the positions of informative polymorphic SNPs. The *solid black dots* and *bars* designate SNPs with allelic loss. (**b**) The genomic map of *RB1* is expanded and shows the positions of the five polymorphic SNPs with allelic loss as well as the positions of two polymorphic DNA markers (RB1.2 and RB1.20) with allelic loss. (**c**) The distribution of clonal allelic losses as they relate to precursor *in situ* lesions and invasive TCC shown as a histologic map at the *bottom* is demonstrated. The *blocks* depict the distribution of clonal allelic losses identified by the hypervariable DNA markers

and was not uniformly associated with the inactivation of the remaining *RB1* allele (Fig. 7.8a–e). In those cases in which deletion of one *RB1* allele was associated with the inactivation of the remaining allele by a point mutation accompanied by loss of RB protein expression, the inactivation was a later event corresponding to onset of HGIN progressing to invasive carcinoma.

Since this region was defined by WOHGM using five cystectomies, we further investigated the frequency of its involvement in 111-paired samples of bladder tumors and peripheral blood DNA by allelotyping of 100 SNPs mapping to a 3.16 Mb segment around *RB1* (Fig. 7.9a–i). The position of the segment with the most frequent LOPs identified by this approach overlapped with the position of the region associated with clonal expansion defined by WOHGM (Fig. 7.9b,c). LOP involving *RB1* and its flanking regions could be detected in over 50% of all bladder cancers and was equally frequent in the two pathogenetic subsets of bladder cancers i.e. low-grade papillary superficial and high-grade invasive TCCs. In contrast, the loss of *RB1* function was predominantly seen in high-grade invasive TCC. These data supported the hypothesis that genes or non-coding sequences located near the model tumor suppressor *RB1*, may contribute to the development of bladder cancer.

The pattern of allelic losses identified by high-resolution SNP mapping across the *RB1* flanking region implied that the breakpoints of the deleted segments were located outside of the coding sequences of the genes (Fig. 7.9c). To examine the genomic context of the minimal deleted region and its associated breakpoints, we performed a comprehensive analysis of the genome sequence characteristics within the 3 Mb segment flanking the *RB1* gene (Fig. 7.9d–i).

Nonallelic homologous recombination (NAHR) is a major mechanism involved in rearrangements during mitotic crossover and DNA break repair causing somatic deletions frequently seen in sporadic human cancer [32,33]. NAHR between similar

---

Fig. 7.7 (continued) (RB1.2, RB1.20, D13S268) and SNPs (*gray blocks*). The code for the histologic map is shown in Fig. 7.2. The hypervariable DNA markers and SNPs with allelic loss associated with plaque-like clonal expansion involving large areas of bladder mucosa were clustered within and around *RB1* and involved approximately 7 Mb. These defined several discontinuous regions of allelic losses associated with early clonal expansion of urothelial cells that ranged in size from approximately 0.27–1.11 Mb and are indicated by the vertical blue bars and *gray shaded areas* in (**a**). The *borders* and predicted size of these regions were defined by the nearest flanking SNPs or microsatellite markers that retained polymorphism. The *numbers* (0.27 Mb, 0.6 Mb, 0.45 Mb, 0.27 Mb, 0.28 Mb, 0.18 Mb, 1.11 Mb) indicate the predicted size of the deleted regions. (**d**) An example of clonal loss of a G/A polymorphism in SNP 6 located within intron 12 of *RB1* is illustrated. Non-tumor DNA of peripheral blood lymphocytes of the same patient (PB DNA) shows G/A polymorphism of SNP 6, while samples corresponding to NU, LGIN, HGIN, and TCC show clonal loss of G. Retention of polymorphism in two SNPs flanking a segment of allelic loss that involves the *RB1* gene is also shown. Overall these data implies that several discontinuous losses of genetic material, which included *RB1* and its flanking regions, occurred in early phases of bladder neoplasia and were associated with *in situ* expansion of a dominant neoplastic clone. (Modified and reprinted with permission from S. Lee et al. *PNAS* 104(34):13732–13737, 2007, panels (**c**) and (**d**) represent new data.)

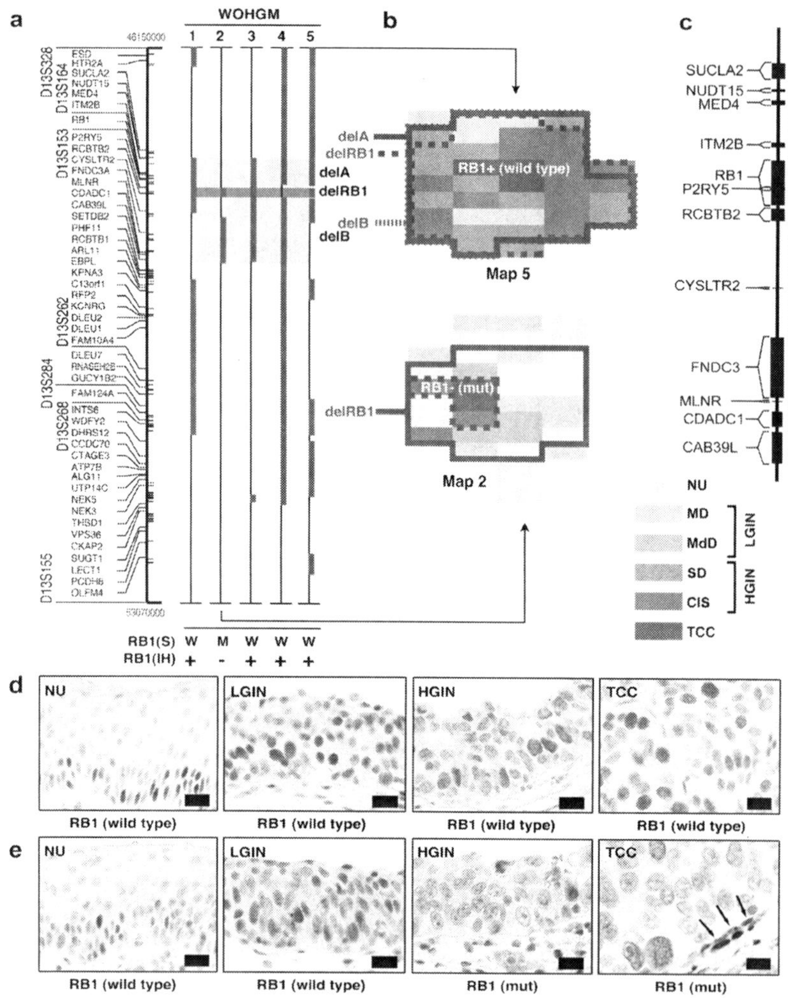

Fig. 7.8 Integration of LOH and LOP patterns identified in the 13q14 region with *RB1* sequencing data and RB protein expression implicating the involvement of FR genes in the intraurothelial expansion of a neoplastic clone. (**a**) Regions of LOP associated with early clonal expansion identified by WOHGM with SNPs in five cystectomy specimens related to the status of *RB1* sequence, *RB1*(S), and RB protein expression revealed by immunohistochemistry, RB(IH), are illustrated. The results of *RB1* sequencing and immunohistochemical studies for RB protein expression are tabulated below the maps of individual bladders. W, wild-type *RB1*. M, mutant *RB1*. The mutation in map 2 involved codon 556 of exon 17 consisting of CGA → TGA and resulting in the change of Arg to a stop codon. The presence of immunohistochemically detectable RB protein is designated by +. The absence of RB protein expression is designated by –, and its distribution pattern is shown in the *lower panel* of (**b**). The genome sequence map, in which the positions of hypervariable markers, as well as known genes are designated by the bars on the left side of map. The regions of LOP in five cystectomies (maps 1–5) are depicted by the *blue solid bars*. The shadowed areas labeled delA and delB designate the regions of LOP flanking *RB1* involved in the incipient expansion of a neoplastic clone. The *shaded area* labeled del*RB1* designates the segment of LOP corresponding to the position of *RB1* on the sequence genome map. (**b**) The distribution of

sequences in the same orientation leads to a reciprocal deletion and duplication between the chromosomes involved. Two large partially overlapping regions bounded by similar sequences were identified. Potential NAHR Region 1 (676 Kbp) is bounded by 474 bp and 429 bp segments that are primate-specific, share 70% similarity and are comprised of ~38% long terminal repeat (LTR) retrotransposons. Potential NAHR Region 2 (1053 Kbp) is bounded by the paralogous genes *RCBTB1* and *RCBTB2*, both of which are candidate tumor suppressors involved in cancer development [34–36]. Potential NAHR Region 3 (67 Kbp) is smaller than the other two regions and is bounded by 750 bp and 745 bp regions with 88% similarity. The positional relationship among the three NAHR regions and deleted segments could be identified in 39% of bladder cancers and involved the vast majority of TCCs demonstrating allelic loss in the 13q14 region. The minimal deleted region flanking *RB1* contained seven recombination hotspots that had a recombination rate higher than the average chromosome 13 recombination rate [37]. The *Alu* density upstream of the minimal deleted region was consistent with the 10.8% genome-wide average, but the regions bordering this peak showed *Alu* densities as high as 39.1% [38].

Human specific retrotransposons such as LINE-1 (L1HS) insertion in *RB1* that has previously been shown to result in germ line deletion of *RB1* exon 24 through non-homologous recombination involving a MER repeat was identified [39]. Human specific *Alu* and SVA insertions were present in the introns of *NUDT15* and *CAB39L* respectively. Human endogenous retrovirus family H (HERVH) insertion in the LOP peak region was also detected [40]. Several structural variants map to the interval, which provides further evidence of genomic instability across the region. These variants were identified in phenotypically normal individuals [41] suggesting that a germ line heterozygous loss of genomic sequence in the region may be associated with an increased risk for bladder cancer. A region of low conservation within the LOP peak between the *RCBTB2* and *CYSLTR2* genes was identified and implies that this region is highly diverse across species suggesting instability over evolutionary time. These analyses suggest several mechanisms that may predispose to losses of genetic material affecting *RB1* and its flanking region.

Fig. 7.8 (continued) clonal LOP involving *RB1* and the same regions shown in (**a**) for map 5 (*upper panel*) is depicted. The *lower panel* shows the distribution of the segment with LOP in map 2 depicted in (**a**). (**c**) Region of clonal LOP associated with growth advantage of *in situ* neoplasia identified by SNP-based mapping. (**d**) The immunohistochemical pattern of RB protein expression in representative mucosal samples of map 5 illustrated in (**b**) as the *upper panel* and corresponding to NU, LGIN, HGIN and TCC is shown. The presence of RB protein in all mucosal samples correlated with the sequencing data, which indicated that the remaining, wild-type *RB1* allele was retained in this case. (**e**) The immunohistochemical pattern of RB protein expression in representative mucosal samples of map 2 illustrated in the *lower panel* of (**b**). Positive nuclear staining for *RB1* protein in stromal endothelial cells serves as an internal positive control (*arrows*). Note the absence of RB protein expression in HGIN and TCC corresponding to an area containing a mutant *RB1* allele. *Solid black bars* within photomicrographs indicate 50 μm. (Reprinted with permission from S. Lee et al. *PNAS* 104(34):13732–13737, 2007.)

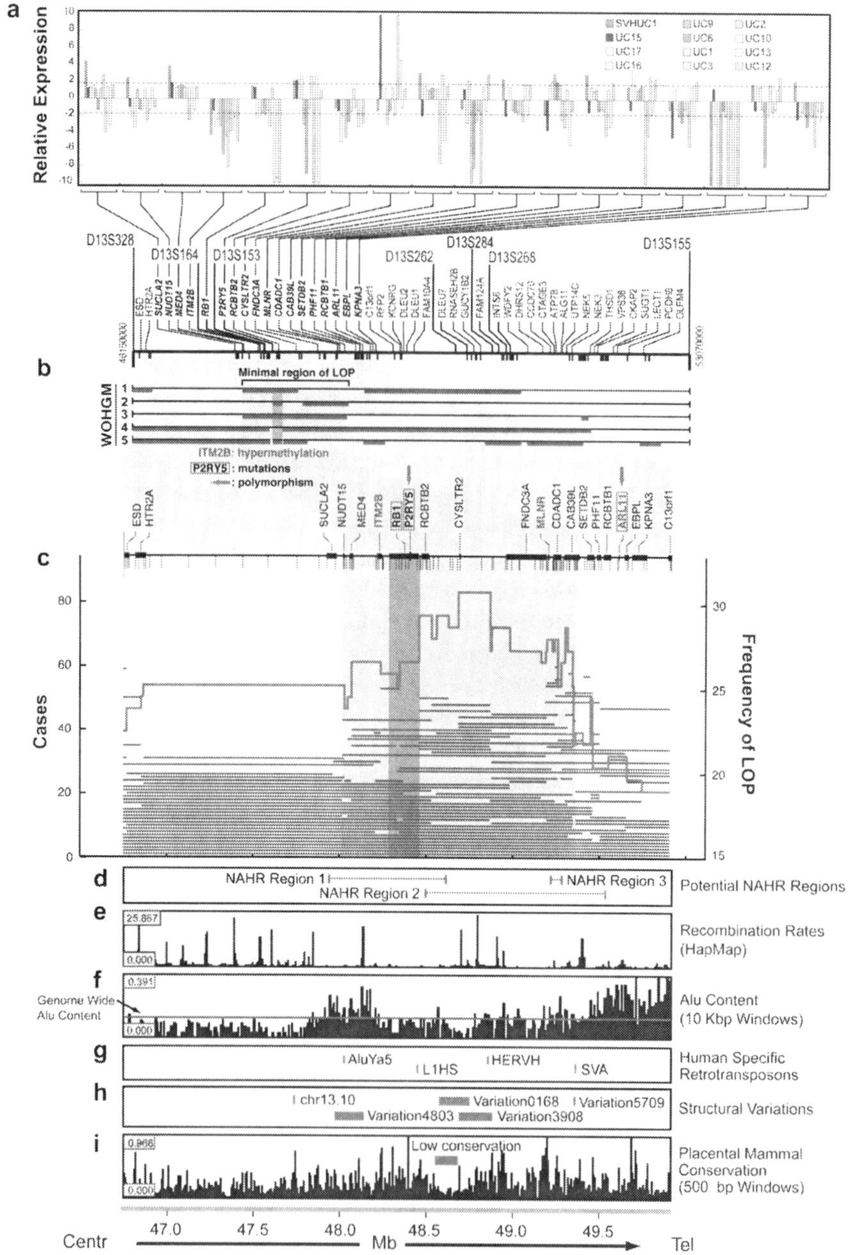

Fig. 7.9 Integration of high resolution SNP-based mapping with expression, epigenetic mapping and genomic content analysis of the 3 Mb segment around *RB1*. (**a**) Results of quantitative RT-PCR showing relative expression of 17 candidate FR genes and *RB1* in 12 bladder cancer cell lines compared to normal urothelial cells. (**b**) Regions of LOP associated with early clonal expansion identified by WOHGM with SNPs in five cystectomy specimens. The pattern of LOP defined a minimal deleted region flanking *RB1* putatively involved in the

Forerunner Genes Contiguous to RB1

Our mapping studies summarized above provided strong correlative evidence supporting the presence of alternative target genes contiguous to *RB1* referred to as FR genes that may drive the initial clonal expansion of *in situ* neoplasia. In fact, the loss of the prototypic FR gene, *P2RY5*, was antecedent to the loss of RB protein expression during tumor development. In this scenario, the silencing of FR genes provided the initial growth advantage for a preneoplastic clone while the subsequent loss of *RB1* function was associated with the development of the successor clone with a fully transformed phenotype i.e. severe dysplasia/carcinoma *in situ* progressing to invasive cancer. Collectively our data suggest that the loss of FR gene function promotes early clonal expansion by presumably directly regulating cell proliferation and/or cell survival [42]. Indeed *ITM2B*, candidate FR gene, flanking *RB1* contained a BH3 domain, a sequence motif that mediates the functions of cell death regulators. Consistent with this hypothesis ectopically driven expression of a wild-type *ITM2B* as well as *P2RY5* reduced rates of proliferation in recipient cells and these effects were associated with induction of apoptosis [15]. The most frequent mechanism responsible for silencing of FR genes was a convergence of allelic loss with methylation of the promoter region. Hypermethylation of the *ITM2B* promoter was identified in 40–50% of bladder cancer cell lines and human tumor samples.

Mutations of FR genes were relatively infrequent and could be identified in less than 10% of bladder tumor samples, but they provided strong evidence supporting the involvement of FR genes in the development of bladder cancer. Missence point mutations in *P2RY5* were found in 7% of TCC and some of these mutations were germ line alterations (Fig. 7.10a). One of the polymorphic sites found in *P2RY5*, (G1722 T) resulting in substitution of cysteine for tryptophan at position 307, was detected in several bladder tumors and non-tumor DNA from the same patient. Molecular modeling of P2RY5 protein suggested that this substitution involving the cytoplasmic domain of the protein might affect its interaction with the G protein complex and compromise its biological activity (Fig. 7.10b). The loss of a wild-type 1722 G *P2RY5* allele with retention of the variant 1722 T *P2RY5* allele occurred in

Fig. 7.9 (continued) development of *in situ* neoplasia. For details see Fig. 7.8. (**c**) LOP tested on 111-paired samples of bladder tumors and peripheral blood using SNP multiplex technology. Predicted sizes of LOP are depicted as *blue bars* and a continuous *red line* shows their frequency. The genomic map above the diagram shows positions of individual genes (*solid black bars*) and tested SNPs (*thin black downward bars*). The pattern and frequency of allelic loss generated by this approach implies that the most frequent breakpoint is located between *RCBTB2* and *CDADC1*. Overall, the pattern of allelic losses suggests the presence of candidate FR genes mapping within the minimal deleted region flanking *RB1*. (**d**) Putative NAHR regions identified by the presence of similar sequences in the same orientation using Human Chained Self Alignment browser. (**e**) Recombination rates based on HapMap. (**f**) Alu repeat content per 10-kb windows. (**g**) Human specific retrotransposons based on UCSC Alignment Nets. (**h**) Human polymorphic structural variants based on the Center for Applied Genomics Database of Genomic Variants and the UCSC Structural Var track. (**i**) Placental mammal conservation scores in 500 bp windows

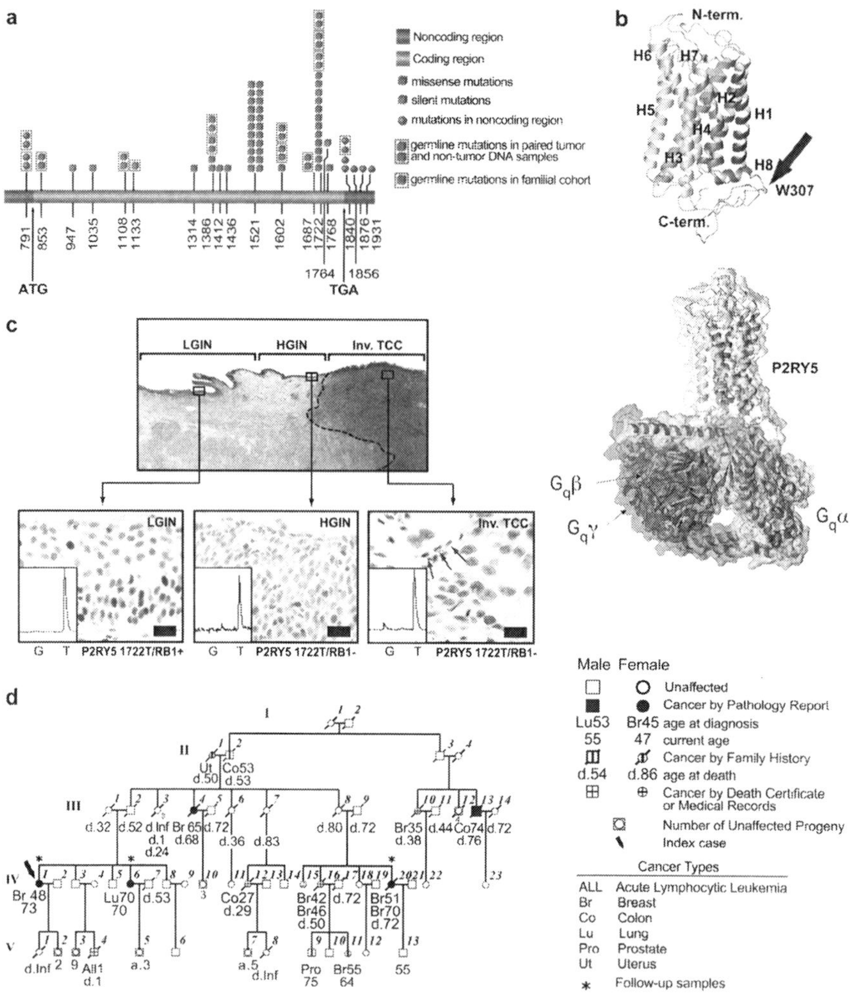

Fig. 7.10 Nucleotide substitutions of *P2RY5* in sporadic and hereditary cancers. (**a**) Summary of sequence analysis of *P2RY5*. The positions of nucleotide substitutions are shown on the full-length mRNA. (**b**) A model of inactive P2RY5 containing 7 transmembrane (H1–H7) and one cytoplasmic (H8) helix structures showing the position of polymorphism in codon 307 located within the cystoplasmic domain of the protein (*top diagram*) that may affect its interaction with the $G_{\alpha\beta\gamma}$ trimeric protein complex (*bottom diagram*). (**c**) Sequential inactivation of *P2RY5* and *RB1* in the development of bladder cancer from *in situ* neoplasia. Low power view of invasive bladder cancer and adjacent LGIN and HGIN. Microdissected DNA corresponding to LGIN shows loss of wild-type *P2RY5* allele and retention of normal RB expression pattern (*left lower panel*). Microdissected DNA corresponding to HGIN shows similar loss of wild-type *P2RY5* allele and additional loss of RB protein expression. Same loss of wild-type *P2RY5* allele and loss of RB protein expression is seen in invasive TCC (*right lower panel*). *Arrows* indicate retention of RB protein expression in endothelial cells adjacent to tumor. (**d**) Pedigree of a family affected by several common human malignancies that include cancers of the breast, lung, colon, prostate, and uterus as well as acute leukemia. Sequencing of the peripheral blood DNA in individual IV1 identified a missence G-C mutation involving codon 111 of *P2RY5*. The same mutation was also detected in individuals IV1, IV6, and IV20. A loss of wild-type *P2RY5* allele and retention of mutant *P2RY5* was identified in breast cancer from individual IV1

the *in situ* phase of bladder neoplasia and preceded the loss of RB protein expression (Fig. 7.10c), and that all 1722 G-T carriers who were smokers developed bladder cancer in a case-control study of 790 patients with bladder cancer. In addition, germ line mutations of *P2RY5* could be identified in families with predisposition for the development of several common epithelial malignancies such as cancers of the breast, colon, lung, prostate, and uterus (Fig. 7.10d). The development of tumors in the members of such families was associated with the loss of the wild-type *P2RY5* and retention of the mutant allele. The initially studied candidate FR genes mapping inside or flanking *RB1, ITM2B, P2RY5*, and *RCBTB2* (*CHC1L*) were down regulated in 63% of cancer cell lines derived from several major groups of common human malignancies indicating their involvement in the development of many cancer types.

In addition to originally studied *ITM2B, P2RY5*, and *RCBTB2* (*CHC1L*), the expression patterns of *GPR38, CAB39L, RCBTB1*, and *ARL11* with more than a 50% reduction of their expression in several bladder cancer cell lines indicate that they should be further investigated as putative FR genes (Fig. 7.9a) [16].

Collectively, our data provide evidence that supports the concept and existence of FR genes. Their loss of function promotes early clonal expansion of *in situ* neoplasia by regulating cell survival via apoptosis and is antecedent to the loss of tumor suppressor such as *RB1* during tumor development [15, 16].

Concluding Remarks

Our studies provide a global look at genome involvement in carcinogenesis and shows that each bladder cancer developed through unique genomic imbalances with a succession of changes providing growth advantage and leading to a gradual change of normal cells into cancer cells [43,44]. Much work remains to be done to produce an accurate high resolution genome-wide map of bladder cancer development, but the significance of this approach and its future applications include the following:

- The landscape of genomic alterations emerging from WOHGM studies implicates cumulative genomic imbalances that occur in early *in situ* phases of bladder neoplasia even before micoscropically recognizable precursor lesions such as dysplasia develop.
- Only a small proportion of genomic imbalances demonstrate an association with clonal expansion of *in situ* preneoplastic lesions and their progression to invasive cancer.
- Genes mapping to the chromosomal regions involved in clonal expansion of *in situ* lesions will help us to understand the molecular mechanisms of the early phases of human carcinogenesis and may represent novel markers for early cancer detection and prevention.
- The analysis of human genome sequences spanning the involved chromosomal regions may provide clues to their structure and instability based on the content of repeat elements, and unique evolutionary features.

- The events of bladder cancer development identified by our WOHGM studies may also be relevant to other pathogenetically related human tumors such as those that arise in the lung and upper aerodigestive tract.

The expanding human genome databases will necessitate a constant redrawing of the picture presented in this review. The high resolution maps of bladder cancer development based on high throughput technologies utilizing allelotyping of SNP's [45] and their correlation with epigenetic and transcriptome maps should provide a more complete picture of cancer development. Such maps in correlation with deeper population sequencing will help point which genomic variants both individually and in combination are involved in the development of cancer.

References

1. Weinberg, R. A., (Ed.) (2006) *The Biology of Cancer*, Garland Science, New York.
2. Fearon, E. R., and Vogelstein, B. (1990) A genetic model for colorectal tumorigenesis, *Cell 61*, 759–767.
3. Hanahan, D., and Weinberg, R. A. (2000) The hallmarks of cancer, *Cell 100*, 57–70.
4. Ross, R. K., Jones, P. A., and Yu, M. C. (1996) Bladder cancer epidemiology and pathogenesis, *Semin Oncol 23*, 536–545.
5. Sidransky, D., and Messing, E. (1992) Molecular genetics and biochemical mechanisms in bladder cancer: oncogenes, tumor suppressor genes, and growth factors, *Urol Clin North Am 19*, 629–639.
6. Simoneau, A. R., and Jones, P. A. (1994) Bladder cancer: the molecular progression to invasive disease, *World J Urol 12*, 89–95.
7. Baselga, J., and Norton, L. (2002) Focus on breast cancer, *Cancer Cell 1*, 319–322.
8. Minna, J. D., Roth, J. A., and Gazdar, A. F. (2002) Focus on lung cancer, *Cancer Cell 1*, 49–52.
9. Isaacs, W., De Marzo, A., and Nelson, W. G. (2002) Focus on prostate cancer, *Cancer Cell 2*, 113–116.
10. Ellenson, L. H., and Wu, T. C. (2004) Focus on endometrial and cervical cancer, *Cancer Cell 5*, 533–538.
11. Carethers, J. M. (1996) The cellular and molecular pathogenesis of colorectal cancer, *Gastroenterol Clin N Am 25*, 737–754.
12. Ilyas, M., Straub, J., Tomlinson, I. P., and Bodmer, W. F. (1999) Genetic pathways in colorectal and other cancers, *Eur J Cancer 35*, 335–351.
13. Watanabe, T., and Muto, T. (2000) Colorectal carcinogenesis based on molecular biology of early colorectal cancer, with special reference to nonpolypoid (superficial) lesions, *World J Surg 24*, 1091–1097.
14. Henson D.E., and Albores-Saavedra, J., (Eds.) (2001) *Pathology of Incipient Neoplesia*, Oxford University Press, New York.
15. Lee, S., Jeong, J., Majewski, T., Scherer, S. E., Kim, M. S., Tuziak, T., Tang, K. S., Baggerly, K., Grossman, H. B., Zhou, J. H., Shen, L., Bondaruk, J., Ahmed, S. S., Samanta, S., Spiess, P., Wu, X., Filipek, S., McConkey, D., Bar-Eli, M., Issa, J. P., Benedict, W. F., and Czerniak, B. (2007) Forerunner genes contiguous to *RB1* contribute to the development of in situ neoplasia, *Proc Natl Acad Sci USA 104*, 13732–13737.
16. Majewski, T., Lee, S., Jeong, J., Yoon, D.-S., Kram, A., Kim, M.-S., Tuziak, T., Bondaruk, J., Lee, S., Park, W.-S., Tang, K. S., Chung, W., Shen, L., Ahmed, S. S., Johnston, D. A., Grossman, H. B., Zhou, J.-H., Harris, R. A., Snyder, C., Filipek, S., Narod, S. A., Watson, P., Lynch, H. T., Gazdar, A., Baggerly, K., Issa, J.-P., Benedict, W. F., Scherer, S. E., Czerniak,

B. (2007) Understanding the development of human bladder cancer by using a whole-organ genomic mapping strategy, *Lab Invest,* 2008.
17. Jemal, A., Siegel, R., Ward, E., Murray, T., Xu, J., and Thun, M. J. (2007) Cancer statistics, 2007, *CA Cancer J Clin 57*, 43–66.
18. Dinney, C. P., McConkey, D. J., Millikan, R. E., Wu, X., Bar-Eli, M., Adam, L., Kamat, A. M., Siefker-Radtke, A. O., Tuziak, T., Sabichi, A. L., Grossman, H. B., Benedict, W. F., and Czerniak, B. (2004) Focus on bladder cancer, *Cancer Cell 6*, 111–116.
19. Spiess, P. E., and Czerniak, B. (2006) Dual-track pathway of bladder carcinogenesis: practical implications, *Arch Pathol Lab Med 130*, 844–852.
20. Koss, L. G., Tiamson, E. M., and Robbins, M. A. (1974) Mapping cancerous and precancerous bladder changes: a study of the urothelium in ten surgically removed bladders, *Jama 227*, 281–286.
21. Koss, L. G., Nakanishi, I., and Freed, S. Z. (1977) Nonpapillary carcinoma in situ and atypical hyperplasia in cancerous bladders: further studies of surgically removed bladders by mapping, *Urology 9*, 442–455.
22. Koss, L. G. (1979) Mapping of the urinary bladder: its impact on the concepts of bladder cancer, *Hum Pathol 10*, 533–548.
23. Collins, F. S. (1995) Positional cloning moves from perditional to traditional, *Nat Genet 9*, 347–350.
24. Collins, F. S. (1992) Positional cloning: let's not call it reverse anymore, *Nat Genet 1*, 3–6.
25. Czerniak, B., Li, L., Chaturvedi, V., Ro, J. Y., Johnston, D. A., Hodges, S., and Benedict, W. F. (2000) Genetic modeling of human urinary bladder carcinogenesis, *Genes Chrom Cancer 27*, 392–402.
26. Chaturvedi, V., Li, L., Hodges, S., Johnston, D., Ro, J. Y., Logothetis, C., von Eschenbach, A. C., Batsakis, J. G., and Czerniak, B. (1997) Superimposed histologic and genetic mapping of chromosome 17 alterations in human urinary bladder neoplasia, *Oncogene 14*, 2059–2070.
27. Czerniak, B., Chaturvedi, V., Li, L., Hodges, S., Johnston, D., Roy, J. Y., Luthra, R., Logothetis, C., von Eschenbach, A. C., Grossman, H. B., Benedict, W. F., and Batsakis, J. G. (1999) Superimposed histologic and genetic mapping of chromosome 9 in progression of human urinary bladder neoplasia: implications for a genetic model of multistep urothelial carcinogenesis and early detection of urinary bladder cancer, *Oncogene 18*, 1185–1196.
28. Yoon, D. S., Li, L., Zhang, R. D., Kram, A., Ro, J. Y., Johnston, D., Grossman, H. B., Scherer, S., and Czerniak, B. (2001) Genetic mapping and DNA sequence-based analysis of deleted regions on chromosome 16 involved in progression of bladder cancer from occult preneoplastic conditions to invasive disease, *Oncogene 20*, 5005–5014.
29. Kram, A., Li, L., Zhang, R. D., Yoon, D. S., Ro, J. Y., Johnston, D., Grossman, H. B., Scherer, S., and Czerniak, B. (2001) Mapping and genome sequence analysis of chromosome 5 regions involved in bladder cancer progression, *Lab Invest 81*, 1039–1048.
30. Tuziak, T., Jeong, J., Majewski, T., Kim, M. S., Steinberg, J., Wang, Z., Yoon, D. S., Kuang, T. C., Baggerly, K., Johnston, D., and Czerniak, B. (2005) High-resolution whole-organ mapping with SNPs and its significance to early events of carcinogenesis, *Lab Invest 85*, 689–701.
31. Kim, M. S., Jeong, J., Majewski, T., Kram, A., Yoon, D. S., Zhang, R. D., Li, J. Z., Ptaszynski, K., Kuang, T. C., Zhou, J. H., Sathyanarayana, U. G., Tuziak, T., Johnston, D. A., Grossman, H. B., Gazdar, A. F., Scherer, S. E., Benedict, W. F., and Czerniak, B. (2006) Evidence for alternative candidate genes near RB1 involved in clonal expansion of in situ urothelial neoplasia, *Lab Invest 86*, 175–190.
32. Bishop J. M., and Weinberg, R. A., (Eds.) (1996) *Molecular Oncology*, Scientific American, New York.
33. Purandare, S. M., and Patel, P. I. (1997) Recombination hot spots and human disease, *Genome Res 7*, 773–786.
34. Mabuchi, H., Fujii, H., Calin, G., Alder, H., Negrini, M., Rassenti, L., Kipps, T. J., Bullrich, F., and Croce, C. M. (2001) Cloning and characterization of CLLD6, CLLD7, and CLLD8, novel

candidate genes for leukemogenesis at chromosome 13q14, a region commonly deleted in B-cell chronic lymphocytic leukemia, *Cancer Res 61*, 2870–2877.
35. Solomou, E. E., Sfikakis, P. P., Kotsi, P., Papaioannou, M., Karali, V., Vervessou, E., Hoffbrand, A. V., and Panayiotidis, P. (2003) 13q deletion in chronic lymphocytic leukemia: characterization of E4.5, a novel chromosome condensation regulator-like guanine nucleotide exchange factor, *Leukemia Lymphoma 44*, 1579–1585.
36. Latil, A., Morant, P., Fournier, G., Mangin, P., Berthon, P., and Cussenot, O. (2002) CHC1-L, a candidate gene for prostate carcinogenesis at 13q14.2, is frequently affected by loss of heterozygosity and underexpressed in human prostate cancer, *Int J Cancer 99*, 689–696.
37. Kong, A., Gudbjartsson, D. F., Sainz, J., Jonsdottir, G. M., Gudjonsson, S. A., Richardsson, B., Sigurdardottir, S., Barnard, J., Hallbeck, B., Masson, G., Shlien, A., Palsson, S. T., Frigge, M. L., Thorgeirsson, T. E., Gulcher, J. R., and Stefansson, K. (2002) A high-resolution recombination map of the human genome, *Nat Genet 31*, 241–247.
38. Grover, D., Mukerji, M., Bhatnagar, P., Kannan, K., and Brahmachari, S. K. (2004) Alu repeat analysis in the complete human genome: trends and variations with respect to genomic composition, *Bioinformatics 20*, 813–817.
39. Albrecht, P., Bode, J., Buiting, K., Prashanth, A. K., and Lohmann, D. R. (2004) Recurrent deletion of a region containing exon 24 of the *RB1* gene caused by non-homologous recombination between a LINE-1HS and MER21B element, *J Med Genet 41*, e122.
40. Yi, J. M., Kim, H. M., and Kim, H. S. (2006) Human endogenous retrovirus HERV-H family in human tissues and cancer cells: expression, identification, and phylogeny, *Cancer Lett 231*, 228–239.
41. Redon, R., Ishikawa, S., Fitch, K. R., Feuk, L., Perry, G. H., Andrews, T. D., Fiegler, H., Shapero, M. H., Carson, A. R., Chen, W., Cho, E. K., Dallaire, S., Freeman, J. L., Gonzalez, J. R., Gratacos, M., Huang, J., Kalaitzopoulos, D., Komura, D., MacDonald, J. R., Marshall, C. R., Mei, R., Montgomery, L., Nishimura, K., Okamura, K., Shen, F., Somerville, M. J., Tchinda, J., Valsesia, A., Woodwark, C., Yang, F., Zhang, J., Zerjal, T., Zhang, J., Armengol, L., Conrad, D. F., Estivill, X., Tyler-Smith, C., Carter, N. P., Aburatani, H., Lee, C., Jones, K. W., Scherer, S. W., and Hurles, M. E. (2006) Global variation in copy number in the human genome, *Nature 444*, 444–454.
42. Fleischer, A., Ayllon, V., Dumoutier, L., Renauld, J. C., and Rebollo, A. (2002) Proapoptotic activity of ITM2B(s), a BH3-only protein induced upon IL-2-deprivation which interacts with Bcl-2, *Oncogene 21*, 3181–3189.
43. Foulds, L., (Ed.) (1969) *Neoplastic Development: The Natural History of Cancer*, Academic Press, New York.
44. Nowell, P. C. (1976) The clonal evolution of tumor cell populations, *Science 194*, 23–28.
45. Peiffer, D. A., Le, J. M., Steemers, F. J., Chang, W., Jenniges, T., Garcia, F., Haden, K., Li, J., Shaw, C. A., Belmont, J., Cheung, S. W., Shen, R. M., Barker, D. L., and Gunderson, K. L. (2006) High-resolution genomic profiling of chromosomal aberrations using Infinium whole-genome genotyping, *Genome Res 16*, 1136–1148.

Index

Printed in the United States
120065LV00003B/72/P

9 781402 083402